THE SUPERNOVA STORY

THE SUPERNOVA STORY

LAURENCE A. MARSCHALL

PLENUM PRESS • NEW YORK AND LONDON

Library of Congress Cataloging in Publication Data

Marschall, Laurence A.
  The supernova story.

  Includes bibliographical references and index.
  1. Supernovae. I. Title.
  QB843.S95   1988   523.8′446   88-17978
  ISBN 0-306-42955-1

89 - B2023

First Printing—August 1988
Second Printing—April 1989

© 1988 Laurence A. Marschall
Plenum Press is a Division of
Plenum Publishing Corporation
233 Spring Street, New York, N.Y. 10013

Printed in the United States of America

To Emma, Geoffrey,
and Ellen

A Blaze in the Sky

The annals of the Benedictine monastery of St. Gallen in Switzerland record the following entry for the year 1006. "A new star of unusual size appeared, glittering in aspect, and dazzling the eyes, causing alarm. . . . It was seen likewise for three months in the inmost limits of the south, beyond all the constellations which are seen in the sky." Chinese court astronomers saw it too, shining with a brilliance that rivaled the moon; several years passed before it faded from sight.

Monks and mandarins had witnessed a rare and wonderful event, not properly a "nova stella" or new star, but the last self-destructive burst of a dying star, an event known to modern science as a supernova. The supernova of 1006 was the brightest in recorded history, one of the very few such explosions to have occurred close enough to our sun to be visible to the naked eye. Fewer than a half-dozen naked-eye supernovae have been recorded since that time, and recent research in the field has largely had to make do with an examination of ancient records, along with analyses of the faint light from supernova explosions in galaxies beyond our Milky Way. Until 1987, not one nearby supernova had been caught *in extremis* by the tools of 20th century astrophysics.

That situation has changed abruptly. Late in the afternoon of February 23, 1987, in a tunnel deep beneath Kamioka, Japan, a dozen flashes of light occurred in a sealed 3000-ton tank of water. No human eye saw them, but all around the tank an army of ultrasensitive photodetectors converted the feeble flashes to faint electrical signals, passing them on to a computer for storage and analysis. Simultaneously, in a salt mine under Lake Erie, a similar apparatus recorded flashes in another giant tank of water. In both cases, the faint glitter marked the passage of a flood of neutrinos, subatomic particles that are released in the decay of the nuclei of atoms. Because a neutrino can pass through a planet like sunlight through glass, even a dozen flashes—a handful of neutrinos snagged in flight—marks a signal of considerable intensity. On February 23, Earth was washed by a great wave of neutrinos, like a torrent from a broken dam. It was the blast from a nearby supernova.

At first neither Japanese nor American physicists were aware of what their apparatus had detected. But by the next night astronomers worldwide were jubilantly passing the word of a brilliant stellar explosion. Ian Shelton, an observer working for the University of Toronto on a mountain in Chile, had photographed a new star, clearly visible to the naked eye, in the southern sky. Located in the Large Magellanic Cloud, a neighbor galaxy to our Milky Way, the exploding star was so bright that when the instruments on some large telescopes were first directed at it, they were saturated with light and unable to measure it properly. One telescope in Chile observed the supernova with its dome partly closed to reduce the level of light falling on its detectors. With all Astronomy in a frenzy over the new discovery, the neutrino physicists checked their data and found, on records made less than a day before Shelton's sighting, the signature of the neutrino pulse.

The burst of light and neutrinos from SN 1987A, named for the date of its arrival at our planet, had been on its journey from the Large Magellanic Cloud for 160,000 years, a long time in the history of humanity. Had the explosion gone off 100 light-years nearer to us, the brilliance would have reached Earth 100 years sooner, arriving in an age when telescopes in space weren't even a dream, when photography was in its infancy, when the analysis of spectra was just beginning to be applied to the study of the sun. Yet, by a great stroke of fortune, the light reached Earth during a time of great technological proficiency, a golden age for astronomy.

Contrast the current sighting with its counterpart a thousand years earlier, and you can appreciate how times have changed. In 1006, the supernova was a blaze in the sky, a portent, a sign from the heavens. Emperors, fearing perfidy, consulted court astrologers; euphoric monks saw visions of the Cross in the sky. The 1987 supernova, considerably farther away, was bright only to astronomers; to the untrained eye it appeared as just one of the many featureless

FIGURE 1: Supernova 1987A is indicated by an arrowhead in this photograph of a portion of the Large Magellanic Cloud taken on February 25, 1987 by Wendy Roberts of the Harvard–Smithsonian Center for Astrophysics using the wide-field Schmidt telescope at Cerro Tololo Inter-American Observatory near La Serena, Chile. The 30 Doradus Nebula, a large gaseous nebula (a gas cloud), is to the upper left. (National Optical Astronomy Observatories.)

specks in the dark night sky. Yet in its own fashion, SN 1987A was as much a public event as the supernova in 1006. It even was accorded a place of honor on the cover of *Time* magazine.

Why all the fuss? Historically speaking, there was nothing new about the 1987 supernova but its name. Medieval and Renaissance astronomers had witnessed such things several times. They called any star that appeared where none had been before a "nova stella," a "new star." Nowadays, astronomers use the old terminology, "nova" for short, but they recognize two sorts: common novae and super-novae. Common novae, as the name implies, are relatively frequent. Supernovae, especially bright supernovae, are exceedingly rare. I will use all three terms. In discussing the early history of astronomy, an era when no one distinguished between common novae and super-novae, I will use the term "nova" to refer to any new star, whatever its nature. In most cases, however, especially where confusion might arise, I shall use "common nova" and "supernova" as appropriate.

Few would have taken notice had the new star of 1987 been a common nova. As we understand them, common novae are explo-sive outbursts that occur among close pairs of stars when one of the pair accumulates material from the other. A critical mass of this sur-face debris may abruptly ignite, giving a distant observer the impres-sion that a new star is appearing from nowhere. An exceedingly bright common nova went off in 1975 among the stars of the con-stellation of Cygnus. At maximum it appeared brighter than the su-pernova of 1987. Yet it received scant mention in the press. Even among astronomers, common novae are something of a curiosity, of interest only to a few specialists. Were supernovae just upscaled nova explosions, they would merit little attention.

Supernovae, their name notwithstanding, represent a phenome-non fundamentally distinct from common novae. Compared to a supernova, a common nova sends out faint sputters of light and feeble puffs of gas. A supernova rends a star to pieces, pushing the laws of nature to the limit. Supernova explosions mark a climactic stage in the lives of stars, and play a central role in the formation of the elements, the shaping of the galaxies, and the evolution of life. Some astronomers suggest that formation of our sun may have been triggered by a supernova, and few doubt that most of the atoms in our bodies originated in the outward rush of matter from one of these exploding stars. Thus, the story of supernovae, ancient and modern, touches on the foundations of biology, physics, cosmology, and astronomy, and draws on the resources of historians, anthropologists, and geologists as well.

In the chapters that follow, we shall trace the threads of that story, from the first sightings of new stars in ancient times, to the latest findings from Supernova 1987A. I shall try to make clear why supernovae generate so much excitement, even though so few of us—astronomers included—have ever seen one with our own eyes. I also hope, along the way, to give you some appreciation of how astronomy is done, and why so many people who do it find it a source of endless amusement and edification.

Laurence A. Marschall

Brookline, Massachusetts

Acknowledgments

Anyone who writes a popular science book inevitably finds it necessary to seek the advice of many others. It could not be otherwise. Science has become so specialized that even in a narrow field like supernova research, which is itself a subfield of astronomy, there are few, if any, who have a total grasp of the entire subject. Even when discussing the same phenomenon, the theoreticians often understand it quite differently from the observers.

I am an observational astronomer—one of those who still use telescopes. My research deals with young stars, not dying ones, though I have had a nonprofessional interest in supernovae for many years. One of the greatest pleasures in writing this book has been discussing supernovae with those whose experience and knowledge is far greater than mine. It has also been a pleasure to work with many others, not just astronomers, who have offered support and criticism, or who have helped obtain illustrations for the book.

Those who have been generous in supplying help and information include Tom Bania, David Devorkin, Oscar Duhalde, Ron Eastman, Andrew Fraknoi, Robert Garrison, Russell Genet, Martha Hazen, Lewis Hobbs, Dorrit Hoffleit, Kenneth Janes, Robert Jedrzejewski, Albert Jones, Steve Kent, Charles Kowal, Donald Lamb, Stephen Maran, Brian Marsden, James Matthews, Richard McCrosky, Peter Nisenson, Bernard Pagel, Costas Papaliolios, Jay Pasachoff, Carl Pennypacker, Rudy Schild, Ronald Schorn, Michael Seeds, Fred Seward, Ian Shelton, Larry Sulak, Victor Thoren, Alan Uomoto, and Nolan Walborn. Special thanks as well to Ken Brecher for many lively lunchtime discussions. And thanks to Jonathan Weiner, who warned me how much labor was involved in writing a book. He was right.

My gratitude also goes to those who helped supply illustrations, among them: Richard Braun, Carolyn Buge, Steve Charlton, Dennis di Cicco, Richard Dreiser, Harold Edgerton, Bruce Elmegreen, Debra Meloy Elmegreen, Margaret Geller, Gus Kayafas, Anne Kottner, Michael Kurtz, Jennifer Mathieu, Mrs. H. Nomoto, Stephen P. Reynolds, Eric Schlegel, Frank Sienkiewicz, Sidney van den Bergh, and Margaret Weems.

I am especially indebted to those who took the time to read and comment on all or part of the manuscript: Ken Croswell, Michael Ferber, John Huchra, Robert Kirshner, Robert Mathieu, Peter Stetson, and Terry Walker. What errors remain, of course, are mine alone. I also am grateful to Gettysburg College for granting me a sabbatical and a leave back-to-back. And I offer my thanks for the hospitality of David Latham at the Harvard–Smithsonian Center for Astrophysics and Alan Marscher at Boston University.

My editors at Plenum Press, Linda Greenspan Regan and Victoria Cherney, ably saw the project through from start to finish; I could not and would not have done the book without them. Finally, my warmest thanks to the members of my family, who have helped this book in so many ways. My parents, Bernice and Frederic Marschall, gave me a love of learning that I hope I will never lose. My father, in addition, took several of the photographs and printed many of the illustrations that appear on these pages. My sister Claudia, who edits, teaches, and directs, has tolerated years of my dubious humor and contributed all of my best jokes. Last, but far from least, I thank my children, Emma and Geoffrey, who exercised extraordinary patience while I typed, and my wife, Ellen Friedman Marschall, who not only read each page, but who labored far harder than I to keep everything going.

Contents

THE SUPERNOVA STORY

CHAPTER 1

A View from Planet Earth

> The universe is made of stories,
> not of atoms.
> —Muriel Rukeyser, *The Speed of Darkness*

STARS AND STORIES

One fine spring day not long ago, in a fit of restlessness, I walked the few blocks from my office to the red brick buildings of the Harvard University Museum. For several hours I ambled through a maze of corridors and halls crammed with specimens, the bounty of a century of collecting zeal. Stuffed animals, large and small, faced me in orderly phalanxes; bones and rocks filled case after case. In one hall, a collection of crystals and gems sparkled under fluorescent lights. In another, a fossil armadillo the size of a Volkswagen loomed against a painted landscape. Tucked away in a small glass case in one room was a plaster replica of the skeleton of "Lucy," childlike in size, yet an adult hominid, one of humankind's earliest known ancestors.

On occasional visits to museums in the past, I'd always been impressed by the diversity of terrestrial life, and the surprising ways in which the simplest physical laws are manifest in nature. This time there was something else. Perhaps because I was setting out to write a book about one of astronomy's rarest phenomena, one that few readers would ever see, I could not suppress a twinge of envy for biologists, paleontologists, and geologists, who, by the nature of their subject matter, could assemble such an eye-catching collection of their work. Here in the museum one could look a lemur in the face, see the layers of sediment that once lay beneath a hundred fathoms of ocean, marvel at the convoluted branches that have sprouted in life's

1

evolutionary tree. Here schoolchildren filed past displays of Gala-
pagos finches, pointed at a brainlike nodule of hematite ore, clustered
around a dinosaur egg. They were directly experiencing some of the
wonder that draws us to science. Admittedly, only a small part of
nature's richness could be locked up in these glass cases, but here at
least were representative items that could be viewed close up, even
touched if one wished, evoking a natural curiosity in the most untrained observer.

The subject matter of astronomy is, by comparison, abstract and
remote. Beyond the sun, the nearest star is so far from us that it
appears as a mere dot in the sky. Even in the great telescopes used by
astronomers, the stars are pinpoints of light, not disks. And larger
celestial objects, the majestic spirals of galaxies, for instance, show
their delicate tracery only on photographs and in video frames taken
by sensitive electronic detectors. A first-time deep-sky observer usually sees little more than a fuzzy glow against the blackness of the
night. Thus, to the nonastronomer, once you've seen one celestial
object, you've pretty much seen them all.

Planetariums, which are the closest things to natural history museums that astronomers have, can produce a reasonable facsimile of
the patterns of the stars and the motions of the planets with ingenious systems of lenses and lights. Beyond that, planetarium direc-
tors must rely on slides of telescopic images or scientifically contrived
artwork to bring the heavens closer to their audience. Not being able
to touch the specimens, many viewers, I suspect, may unconsciously
regard the picture of the universe presented by modern science as no
more real than a space fantasy from Hollywood. If only the fascinating array of celestial objects could be put into glass cases. If only we
could view a star or nebula close up.

Yet in the balance, astronomical research deals with concepts no
more fanciful or remote than paleontology or geology. Astronomy's
bizarre species and evolutionary forms lie hidden above our heads,
just as gems and bones lie buried beneath our feet. Sensitive detectors
and sophisticated methods of analyzing light serve astronomers
much as the excavating tools and camel's hair brushes serve the pal-
eontologist. In the field, both the galaxy and the fossil are equally
undistinguished to the casual observer, and an exploding star can be
as elusive as a rare tropical insect. Thus, to look at a meticulously
reproduced astronomical photograph is to look at a cleaned and

mounted specimen in a museum case. Scientific acumen, technical skill, and a little good luck are essential ingredients of both.

There is an even more fundamental kinship between the sciences of the Earth and of the sky. In astronomy, as in other branches of science, a specimen is incomplete without a story. Darwin's finches would be nothing more than feathered ornaments without the rich web of history that links them to the theory of natural selection and, by implication, to the origin of our species. The most flamboyant crystals would be nothing more than geometric curiosities without the knowledge that their regular shapes and striking colors are a consequence of the forces that bind their atoms together.

And a supernova would be little more than a spark in the sky without a realization of the importance of such stellar explosions to our understanding of the structure and history of the universe. Supernovae, among other things, play a crucial role in the life cycles of stars and in the formation of the chemical elements. Supernovae scatter the elements of life among the stars and, it is thought, trigger the formation of new stars and planetary systems. It is even possible, though far from certain, that radiation from supernovae causes the mutation of the genetic material in living organisms, thereby providing a mechanism for evolutionary change.

The story of supernovae, remote and rare as they are, is intimately related to the story of the ordinary stars of the nighttime sky, and cannot be appreciated without some overview of what we know about them and how we came to know it. The story of the stars is a grand and expansive tale, and we can only give the briefest sketch of it here; astronomy textbooks contain many details that need not concern us. But some knowledge of astronomy, of how normal stars appear to the naked eye, of how they compare to our own star, the sun, and how they are arranged in space, will help us understand the significance of stellar explosions in the overall scheme of things.

In the overall scheme of things, in fact, the supernova story is ultimately related to the specimens we see, row upon row, in the great natural history collections of the world. The finch, the crystal, and the distant star are all part of a system of nature that has only gradually been understood through millennia of observation and speculation. Though we shall keep our eyes on the stars and tell, as best we can, only stellar tales, we should keep in mind that there are always underlying connections between the remote and the familiar, between the barely discernible depths of space and the intricate living planet of which we are a part.

WHAT ASTRONOMERS DO

The story of the stars is a story in two senses. It is first a history of discoveries, a story told *about* astronomers and the things they have observed. Since its beginnings, the astronomer's role has changed profoundly. The earliest astronomers, we shall see, were celestial lookouts, scanning the skies like seamen in a crow's nest. Early astronomy was intertwined with astrology, and ancient stargazers, particularly in China, watched assiduously for portents of the future displayed in the sky. They discovered much that remains of value today. Later astronomers were assiduous ledger keepers, tracking and recording the incredible variety of sights in the heavens, gradually learning to distinguish one type of object from another, slowly reaching an understanding of what they were looking at. In modern times astronomers have become physicists, extending our knowledge of terrestrial phenomena to the boundaries of the universe.

The story we shall tell is a story in a second sense: it is a story told *by* astronomers. The ledgers are kept and the discoveries valued, by and large, for their power to explain. Astronomers, like scientists of all persuasions, are concerned with understanding and relating the things they see to an overall scheme of nature. Their stories are interpretations of the view from Earth, attempts to go in the mind's eye where they cannot go in person.

You have only to attend a meeting of professional astronomers to realize that this is the case. The language may at first seem unintelligible and the subject matter recondite, but there's a familiar narrative structure to each presentation. Here's a hypothetical example: A curious red star appears to show regular variations in light. After the slides have been shown and the figures listed on the blackboard, the astronomer pauses, looks at the assembled colleagues, and says, in effect, "This is what I think we're seeing. What looks like one star is really two, in orbit around each other. The fainter star periodically blocks out the light of the brighter." Another listens and objects, "Couldn't this explain it more effectively? A single star is swelling and shrinking. When it is larger, it appears brighter; when it is smaller, it appears fainter." And another counters: "I don't think so. Here's what's wrong with your explanation. . . ."

The account is a collective one, like a tale told around a campfire with each person in the circle adding an episode. If you listen to a paper at a scientific meeting, or read a chapter at random from this book, you may get the impression that astronomical stories are just fragments, bits and pieces about a peculiar star here, a distorted galaxy there, a puzzling wisp in a nebula somewhere else. Hearing just a part, you may fail to appreciate what has gone before, missing the overall direction of the enterprise. The ultimate aim is a smoothly interlocking account: a collaborative effort, but a unified tale.

However, there are crucial differences between scientific story-telling and campfire yarn-spinning. A scientific account is one formed over the generations. Its parts can be written and rewritten, and even the entire story line changed, as more information, better theories, and more clever storytellers enter the circle. Unlike a campfire story, cleverness or wit won't guarantee acceptance. Astronomical accounts must agree with observations, successfully predict the future, and fit in agreeably with what is already known and accepted. Clarity and consistency count for everything.

An astronomical account of a celestial phenomenon follows a stock story line: the astronomer presents observations, poses questions, and offers explanations. Some of the questions seem surprisingly simple. How far away is that glowing object we see in the sky? What would it look like if we could see it close up, if we could touch it, weigh it in our hands, assay its chemical makeup? Simple questions. Yet, as we shall see, these fundamental problems are often vexingly difficult to answer.

Even more difficult are questions of structure and origin. What makes a star work? Where does a star come from? How long is its gestation, its childhood, its maturity? What are its kin among the other objects we see above us?

And ultimately there are the large-scale questions: How does this object fit into the entire cosmic order? How is it connected with everything else we know about the universe?

All of these elements of astronomical narrative are present in the scientific account of supernovae. Recent observations of Supernova 1987A in the Large Magellanic Cloud have given astronomers confidence that many of the features of the accepted story are fundamen-

tally sound, particularly our understanding of supernovae as the collapse of very massive stars.

But you should not get the impression that the story of supernovae is anywhere near complete. Because it has been rewritten and
revised many times, because its chapters have been deleted and re-
titled and broken up and reassembled more often than we can count,
the story of supernovae, like all good scientific stories, must be seen
as an unfolding, an organic body of knowledge that has grown and
adapted over the generations, and will continue to do so in the
future.

LIGHTS IN THE SKY

In the several millennia of recorded history, we have only recently realized the simple fact that not all stars shine with a constant,
eternal light. It was not that exploding stars were not seen; there are
well-confirmed historical records of at least seven supernova explosions during the last 2000 years. It was rather that such phenomena
were seldom distinguished from the many other kinds of changes
that can be seen in the night sky. Among these short-term visitors are
meteors, halos around the moon, odd-shaped clouds, aurorae, comets, variable stars (stars that fluctuate in brightness in either a regular
or an irregular fashion), and "new" stars or novae (stars that suddenly appear and later fade and vanish).

To our modern sensibilities, this is a rather diverse group of
objects. Some are clearly atmospheric, like halos around the moon
(which are produced by ice crystals scattering moonlight). Others,
such as comets, variable stars, and novae, are clearly celestial. And
some are a combination of both. Meteors are seen when celestial
debris, usually sand-size particles of interplanetary material, collide
with Earth and are heated to incandescence by friction with the air.
Aurorae result when gas atoms in the upper atmosphere are bombarded by energetic cosmic rays from the sun; the atoms glow much
as a tube of neon in an advertising sign glows when an electric current is passed through it.

That we make these distinctions with little effort, is a consequence of generations of scientific observation. Unlike our ancestors,
we can fly among the clouds, sample the remains of meteors, and
analyze starlight with spectrographs. It is too easy to forget that the

simple physical distinction between atmospheric and cosmic, between nearby flickerings and gargantuan catastrophes, was only
clarified in the last few centuries. Before that, virtually anything that
brightened in the heavens was a "new" light. The interpretation put
on it depended more on the cultural or philosophical background of
the observers than on any quantitative measurement of its properties.
Chinese astronomers, for millennia, viewed the various celestial
changes as portents and omens, while medieval European scholars,
entranced by a notion of the heavens as divine and changeless,
denied the possibility of any cosmic changes whatsoever, and regarded all alterations in the sky as mere atmospheric disturbances.

Even when "new stars" were recognized as a fundamentally distinct phenomenon in the 17th and 18th centuries, even when it was
conjectured that stars could explode, the distinction between the
common novae and the far more powerful supernovae, was long in
coming. "Nova" was simply the term applied to any starlike object
that brightened radically in a short period of time. Although observations of supernovae date back to at least the 2nd century A.D. (and
probably much earlier), the notion of a supernova itself is only 50
years old. Part of the supernova story, then, has been a growing
sense of distinctions, an astronomical coming of age.

TAKING STOCK OF THE SKY

Given the remoteness of the heavens, it is not really surprising
that a sense of astronomical distinctions was slow in coming.
Changes in the heavens are infrequent enough, and subtle enough,
that a single observer would find it difficult to amass a sufficient stock
of experience to judge what has changed and what has not. The
immediate impression of the heavens, in fact, is one of permanence
and predictability, a feature that, as far as we can tell, was central to
the beliefs of the earliest literate societies, the Babylonians, Egyp-
tians, and Chinese, and later the Greeks.

The most fundamental observation that they must have made
was that the patterns of most of the stars remain unchanged. By the
first millennium B.C., both the Chinese and the Greeks represented
the heavens as a sphere to which the stars were attached. Once a day
the rotating sphere carried the stars around the heavens. The stars
moved, but not with respect to each other: groupings of stars, which

were named after familiar heroes, animals, and objects, had retained their distinctive forms for as long as anyone could remember. Except for their twinkling, the stars did not vary noticeably in brightness either.

It must have been readily apparent that the moon was an exception to the rule. It did not remain fixed among the stars, nor did it remain fixed in brightness. Every 30 days it circled once around the heavens, changing phase as it did so. The lunar cycle provided an important measure of time in most societies: the start of a new month could easily be determined by watching for the reappearance of the new crescent moon. In Chinese astronomy, the moon's motion through the constellations also provided a series of landmarks in the sky. The celestial sphere was divided into 28 "lunar mansions," of varying width, each identified with a distinctive grouping of stars that lay along the moon's monthly track.

The ancients also realized that the sun moved around the sky just as the moon did. They could not see the stars during the daytime, of course. But the constellations visible immediately after sunset and immediately before dawn changed slowly as the sun traced and retraced an annual circle, the ecliptic, from west to east around the sky. The regularity of this motion formed the basis of another important measure of time: the year. Early astronomy thus owed much to the need for a reliable calendar. Planting and harvest could easily be measured by the appearance of a key star or constellation in the morning or evening sky.

The only other moving objects regularly seen in the sky were what we know today as the five brightest planets—Mercury, Venus, Mars, Jupiter, and Saturn. Their motions were more inscrutable, for they did not cross the sky at the same rate, sometimes reversed themselves (we call this east-to-west motion *retrograde*), and often changed brightness over the course of the months. Both the Chinese and the Babylonians regarded the changing relationships among the planets as important portents of affairs on Earth, and watched with care for particularly auspicious or ominous arrangements.

In these early times, the distinction between planets and stars was not what it is today. Planets, to us, are bodies that circle the sun, shining with reflected sunlight. Stars are distant suns, shining because they are hot and incandescent. But the ancients regarded the planets simply as movable stars. Our word *planet*, in fact, comes from the Greek word for a "wanderer." Thus, the sun and moon, along with Mercury, Venus, Mars, Jupiter, and Saturn were planets. Aside from their tendency to wander, they were not constitutionally different from the stars in the constellations.

The overall fixity of the constellations, despite the night-to-night turning of the celestial sphere, made it possible to locate events in the heavens by referring them to nearby constellations or asterisms (close groupings of a small number of stars). By the 3rd century B.C., Chinese astronomers had catalogued over 800 stars in more than 125 constellations, and had constructed the first useful maps of the heavens. Thus, in a Chinese text we may read a typical passage: "When Mars is retrograding in the lunar mansion Ying-shih (which includes parts of the modern constellations of Pisces and Pegasus), the ministers conspire and the soldiers revolt."

Unusual events, such as the appearance of a bright meteor or a comet, were recorded in this fashion as well. Today we recognize that each star has its own slow motion with respect to the others in the heavens, called its proper motion (to distinguish it from the daily collective rotation of the heavens), and we are aware that the shapes of the constellations do change with time. Nevertheless, on the scale of human history, the changes are slight. The ancient records are still intelligible two and three thousand years after they were recorded and, as we shall discuss later, still provide valuable historical information on a wide variety of celestial phenomena, from comets to supernovae.

Subtle changes in the heavens occur all the time. Under the scrutiny of modern astrophysical instruments, virtually every star alters position with respect to its neighbors; variations in brightness are also common, if not ubiquitous. But only the most spectacular events caught the attention of the ancient observers. An eclipse of the sun or moon, of course, could not be ignored, nor could a bright comet, trailing a luminous tail for tens of degrees across the sky. Such events might be recorded a dozen times during a single observer's lifetime.

Exploding stars, however, are in a class by themselves. The complication is that an average star, suddenly bursting forth in a flare of light, is bound to get lost in the crowded heavens unless it reaches exceptional brilliance. As we noted earlier, a star, even an exploding one, appears as an indistinguishable point, like all the others in the sky. About 9000 stars can be seen without the aid of a telescope, and though only about half of these are above the horizon at any one time, a total recall of the entire sky, from the faintest speck to the brightest spark, is far beyond ordinary powers of memory. A few times each millennium, the sudden intrusion of a new and extremely bright star in a constellation conspicuous and familiar to many observers would attract widespread attention. A more subtle change was bound to go unnoticed.

It was just such an event that caught the attention of the Greek astronomer Hipparchus of Nicea in the 2nd century B.C., motivating him, so the story goes, to compile the first catalogue of fixed stars and their positions (expressed in degrees of latitude and longitude) with respect to the ecliptic. The claim that Hipparchus's celestial list was sparked by an exploding star rests on dubious evidence. We have only the word of the Roman author Pliny (c. 70 A.D.) that Hipparchus had "discovered a new star, and another one that originated at that time." What does this mean? Some argue that it is a reference to a nova, but there are no other historical records of a "new" star that can be identified with it. Fact or legend, it seems fitting that a stellar explosion should have motivated one of the first systematic surveys of the heavens.

Hipparchus's original catalog of 850 stars does not survive, but it was essential to the development of a science of astronomy in the West. The catalog was adopted by the second-century Greek astronomer Claudius Ptolemy, who added 170 more stars and published it as Volume 7 of his influential textbook of the heavens known as the *Almagest*. For almost 1500 years Ptolemy's compendium remained the standard against which all astronomical treatises were measured, and his catalog was a fundamental reference for astronomical observations and for navigation.

Each entry in the catalog identified a star by its position in a constellation (e.g., in the head of the Bull, in the leg of Orion), by its latitude and longitude in the sky, and by its magnitude, a measure of its brightness. On Ptolemy's magnitude scale, the brightest stars in the sky were classified as magnitude 1, the next brightest magnitude 2, and so on down to magnitude 6, the faintest that could be seen. Modern astronomers have adopted the ancient magnitude scale with some refinements, so that very bright objects have negative magnitudes (the sun is magnitude -26.3 and Sirius is -1.5), and very faint objects, visible only through a telescope, have magnitudes ranging into the 20s. The faintest objects that can be detected with the largest optical telescopes, at about magnitude 25 or 26, are 100 million times fainter than the faintest stars Ptolemy could discern with the naked eye.

The star catalogs of Ptolemy and Hipparchus served astronomers in two ways. First, they established a network of reference points, the 1020 stars, for accurately recording the positions of celestial events. This was the principal work of astronomers at that time, since prior to 1609, when the telescope was first turned toward the heavens, astronomers could study the sky only with the naked eye, aided by sighting devices to measure the angles between stars. Using a standard catalog, astronomers could record not simply that Mars was bright in the southwest, but that Mars was one degree east of a certain reddish star in the eye of Taurus, and that the planet was as bright as a star of magnitude 2.

Equally important, the catalogs opened up the possibility of seeing the heavens in historical perspective. (An even greater number of ancient Chinese records permitted this as well, but they were not used for this purpose until the mid-19th century.) Given a written catalog and sufficient time, minute changes in the heavens became detectible. Without a catalog, astronomers' memories were the only record keepers; if a star moved slightly, or faded a bit over a century or so, it went unnoticed. The memory of individuals was too limited to realize that something had changed. But with the advent of celestial records it was possible to note alterations in the sky with precision, and to preserve the records for later generations.

Of course it was far easier to recognize the intrusion of odd-shaped objects than the appearance or disappearance of stars. Halley's Comet is the most famous case in point. For centuries its return (approximately every 75 years) was recorded in both Europe and China. (The earliest confirmed Chinese sighting dates to 239 B.C.) It was regarded as an omen of defeat for the Saxon King Harald at the Battle of Hastings during its appearance in 1066. It was duly noted in Europe in 1531, 1607, and 1682. It is noteworthy, however, that each time it was regarded as a different object. Not until 1705, when Edmond Halley, a contemporary of Isaac Newton, analyzed the orbits of 24 well-charted comets, did it become clear that one comet was returning with predictable regularity. The return of the comet in 1758, in accord with Halley's calculations, assured him and his comet a place in popular history.

It was also Halley, to his lasting credit, who first used the ancient catalogs to note that the "fixed" stars were not really stationary. Writing in 1718, Halley noted that Arcturus and Sirius, two of the most brilliant stars in the heavens, were no longer in the precise positions given by the old Greek catalogs. The differences were considerable, about a half a degree, the diameter of the full moon in the sky. It was inconceivable that Hipparchus or Ptolemy could have made so glaring an error, even with the relatively crude instruments of their day. Rather, reasoned Halley, it seemed likely that the stars themselves had moved. "What shall we say then?" he wrote. "These stars being the most conspicuous in Heaven are in all probability the nearest to the earth; and if they have any particular motion of their own, it is most likely to be perceived in them." He was correct; they had moved. (However, of the two stars, only Sirius is among the nearest to the Earth.) Note, though, that a long time had to pass before the change became noticeable—the rate Halley calculated was so slow that in the fifty-year career of an astronomer, one of these relatively speedy stars would shift only one minute of arc (1/60 of a degree) in the sky, about the diameter of a dime as viewed from the far side of a football field. Little wonder that the old star catalogs had sufficed for such a long time.

If the stars could wander, could they also vary in brightness? The old catalogs, it is true, listed magnitudes as well as positions. Estimates of magnitude, however, were subjective and difficult to verify, so it is not surprising that the discovery that stars could vary in brightness was slow in coming. Of course there were appearances of brilliant new stars whose presence could not be overlooked. The novae of 1572 and 1604, which we shall discuss in some detail in Chapter 4, were as bright as the brightest stars in the heavens. But fainter novae went unnoticed. David Clark and F. Richard Stephenson, who have studied the ancient records meticulously, estimate that a nova would have to brighten above magnitude 3—about as bright as the stars of the Little Dipper—to have a chance of being recognized by pretelescopic astronomers. A few observers in the 16th and 17th centuries reported fluctuations in brightness of some stars, but it was not until the 18th century that astronomers recognized that there was a distinct class of stars that brightened and dimmed, some in a regular and predictable fashion, others only sporadically.

Halley's discovery of stellar motions established the value of careful cataloging of the heavens and led to a far greater understanding of the position and brightness changes of the stars. His work came at a time when major improvements were being made in the precision with which astronomical instruments could measure star positions. The telescope had been introduced into astronomy by Galileo a century earlier (in 1609) and telescopic sighting instruments with carefully machined scales were replacing the old methods of position measurement, which had relied exclusively on the naked eye. At the same time, a growing commercial navy, sailing halfway around the world to colonize and to trade, was demanding astronomical tables to aid in navigation. All over Europe, astronomers were led more and more to the task of systematic position measuring, brightness estimation, and ledger keeping. Through the 19th century, in fact, the main task of astronomers was the meticulous mapping of the heavens and the cataloging of their contents.

That was no mean task. Ptolemy's compilation contained 1020 stars, including all the bright stars we recognize in the prominent northern constellations. It was not replaced with a more definitive catalog until the publication, in 1725, of the *Historia Coelestis Britannica*, the life work of the first British Astronomer Royal, John Flamsteed. Flamsteed's catalog contained 3000 stars, among them, as we shall note later, an object that is no longer visible today—possibly a supernova. All Flamsteed's stars were brighter than the limit at which a star can be seen by the naked eye. But as telescopes revealed more and more faint stars just a bit fainter than this limit, the goal of exhaustive mapping receded steadily into impossibility.

If this seems difficult to accept, consider the following figures. There are about 9000 stars visible to the naked eye—brighter than about magnitude 6.5. Going a mere 10 times fainter, to magnitude 9, adds another quarter of a million. Counting that many stars by eye, even on photographs of the sky, was virtually a lifetime task for an entire observatory of astronomers. Several such catalogs have been produced over the years. Yet by modern standards, 9th-magnitude stars are considered relatively bright. Using photography and computerized measuring machines, astronomers have recently undertaken to compile a complete list of celestial objects down to magnitude 15 (about 4000 times fainter than the naked-eye limit) to serve as a guide to finding objects using the Hubble Space Telescope (scheduled for launch on the Space Shuttle). That catalog, whose faintest stars are nonetheless almost a million times brighter than the faintest objects detectable with the Space Telescope, contains about 20 million entries. (This, incidentally, points to a major difficulty in discovering supernovae by indiscriminately mapping and remapping the sky. Unless supernovae are very frequent, searching for an occasional interloper in a catalog of 20 million stars is literally more difficult than finding a needle in a haystack.)

But if the celestial surveys of the 18th and 19th centuries were incapable of achieving the ideal of complete coverage of the heavens, they paid off handsomely in collecting exotic specimens that led to an understanding of the system of stars in which we reside. One of the pioneers in this endeavor was the French astronomer Charles Messier, whose principal interest was in discovering comets. These were usually easily recognizable through a telescope because of their hazy or nebulous appearance, and because they appeared to move among the stars during the course of an evening's observation.

In 1771, Messier published a list of a hundred nebulae (singular "nebula"). These were hazy objects that did *not* move. Messier's intention was to preclude their confusion with comets. He himself had been embarrassed by announcing, in 1758, that one of these nebulae was the long-awaited comet of Halley, and he was anxious not to repeat the mistake. The Messier catalog of nebulae has since become a standard listing of "extended" (i.e., nonstellar) objects in the sky.

What the nebulous objects really were was unknown at the time; there was no way for Messier to measure their distances. Among them, we know today, are relatively nearby clouds of gas, small clusters of stars, and vast, distant galaxies that rank among the largest bodies in the universe. First on the Messier list (and therefore known to astronomers as M1) was a ragged cloud of gas in the constellation of Taurus. Called the Crab Nebula, it is the expanding debris from one of the most recent supernova explosions witnessed in our galaxy.

Another fruitful survey of the stars was conducted in the late 1700s by a British amateur, William Herschel. Assisted by his sister Caroline, he systematically searched the sky with large telescopes of his own construction, counting and recording whatever he saw. In addition to cataloging several thousand clusters and nebulae (and discovering Uranus), Herschel's survey marked the first attempt to determine the size and extent of the star system to which our sun belongs, the Milky Way Galaxy. Herschel merely pointed his telescope in a particular direction and counted all the stars he could see through the eyepiece. By assuming, quite incorrectly it turned out, that all stars give off the same amount of light, Herschel translated his star counts into a measure of how far the Milky Way extended in that direction. Herschel surveyed over 3000 directions in this fashion and concluded, in 1785, that the Milky Way was a flattened disk rather like a hockey puck—Herschel referred to it as a "grindstone." The grindstone of stars was about five times as broad as it was thick, with the sun located nearly at the center. Herschel's view of the Milky Way was limited, in retrospect, by his imprecise knowledge of the properties of stars and the nature of the space between them. We are aware today of clouds of gas and dust that hide the vast expanse of the Milky Way from easy view and that truncated Herschel's counts short of the true edge of our galaxy. Yet even so, Herschel's vision of the "stupendous sidereal system we inhabit . . . consisting of many millions of stars" emphasized a feature that will raise little argument among moderns: the sun occupies but a small part of the vast expanse of space.

As position-measuring techniques became more and more precise, it eventually became possible to gauge the distances to the stars more directly. Seen from opposite sides of Earth's orbit (i.e., using observations made six months apart), nearby stars show a shift in direction, called annual parallax, caused by the change in perspective of the observer. The size of the shift is a measure of the distance of the star: the more distant a star, the smaller the parallax. Though astronomers were long aware of the principle of parallax, it was not until 1838 that astronomical measurements became precise enough to measure it.

In that year, the German astronomer Friedrich Wilhelm Bessel announced measurements of a 5th-magnitude star in the constellation of Cygnus (61 Cygni in Flamsteed's catalog). Its semiyearly shift was about a third of a second of arc (1 second = 1/3600 of a degree), an angle that corresponds to the size of a dime viewed from a distance of about four miles. That translated into a distance to 61 Cygni of almost 600,000 times the distance to the sun. Today the method has been applied successfully to about 10,000 stars, ranging out to distances almost thirty times as great.

Though distances such as this are common in astronomy, they strain our powers of comprehension. 61 Cygni is over 50 trillion miles away. The light measured by Bessel, traveling a million miles every 5 seconds, took almost 11 years to make the journey from the star to his telescope: we say that 61 Cygni is nearly 11 light-years from the sun. Yet among the several hundred billion stars that populate the Milky Way (according to current estimates), 61 Cygni ranks among the few

FIGURE 2: William Herschel. The paper he is holding announces his discovery of the planet Uranus, which he wished to name after King George III. (Yerkes Observatory.)

dozen closest to us. Clearly Herschel was correct, perhaps even conservative, in describing our stellar system as "stupendous."

For all the measurement and tabulation, however, the stars remained remote and mysterious. By the mid-19th century it was clear that they were distant suns, some a bit fainter than our own, others many times more luminous. There the information stopped. How could one hope to understand more about the stars, when the very distances between them were, for all practical purposes, insurmountable? In an age when natural history was a popular avocation, when naturalists were beginning to amass their vast collections of animal,

FIGURE 3: Caroline Herschel. (Yerkes Observatory.)

vegetable, and mineral specimens, astronomy seemed to have reached a critical impasse. Was there nothing left to do but to make longer and longer lists?

Typical of the situation was an oft-quoted comment by the French social philosopher Auguste Comte. Writing in 1835, Comte claimed that the true nature of the stars would remain forever obscure to earthbound man. "In a word, our positive knowledge with respect to the stars is necessarily limited solely to geometric and mechanical phenomena."

How wrong he was. Unknown to Comte at the time, the German physicist Joseph von Fraunhofer had, almost 20 years earlier, ob-

served dark lines crossing the spectrum, or rainbow, of color produced when sunlight was passed through a glass prism. That pattern
of dark lines, whose origin was then unknown, signaled the end of
our ignorance of the true nature of the stars.

In the mid-1800s Fraunhofer's compatriot Gustav Kirchhoff, and
his colleague Robert Bunsen at Heidelberg, showed that some of
Fraunhofer's lines were caused by gas atoms that absorbed specific
colors of light, thus removing those colors from the beam of sunlight.
It was not immediately clear whether the absorption took place in the
atmosphere of Earth, or on the surface of the sun itself. (Both were
true.) But by the end of the century the spectroscope, a device for
breaking up starlight as Fraunhofer had broken up sunlight, had
become a standard astronomical tool, and a wide range of techniques
had been developed to wring information from the distinctive patterns of light and dark in stellar spectra.

Spectral analysis of radiation from the stars proved the key to
understanding the physical makeup of the stars and other objects that
populate our universe. Temperatures, chemical compositions, sizes,
and even ages of stars eventually became open for inspection.
Astronomers began to unlock the secrets of how stars were born, how
they burned, and why they sometimes exploded. Spectral analysis
led to the charting of countless galaxies beyond the Milky Way and to
the discovery that the universe is expanding. As astronomers became
adept at analyzing light, they began to develop new techniques of
astronomy that utilized radiation beyond the ends of the visible spectrum: infrared, X-ray, and radio astronomy, among others. Fraunhofer's discovery marked the birth of astrophysics, moving astronomy out of the realm of meticulous ledger keeping into the narrative
mainstream of science. Since his time, astronomical knowledge has
developed at a remarkable pace, all the more remarkable because it
derives so much from so little. It is a story written in the subtle
language of starlight.

CHAPTER 2

Messages in Starlight

Stars scribble on our eyes the frosty sagas,
The gleaming cantos of unvanquished space.
—Hart Crane, *Cape Hatteras*

Gaze at a distant shore on a clear night and you can appreciate the eloquence of starlight. Pinpoints of light glisten on the horizon, as featureless as the stars. Yet you can identify them easily by their brightness, color, and behavior. A momentary flash of brilliant white comes from the lighthouse on the point; a regular red pulsation marks the radio tower; an indistinct glow comes from the porch lights of a community of beach houses; and a garish yellow shines from the parking lot by the pier. Starlight, like shore light, can be recognized by its salient features, once we have learned what to look for.

THE MEASURE OF THE STARS

The simplest distinction we make is between bright and dim lights, of course. All the houses we see on a shore are roughly the same distance from us. Thus, the brightest lights come from the most powerful lamps. When we look at the sky, however, we are seeing stars at various distances from us. If all stars gave off the same amount of light, if they were equally luminous, to use the technical term, the more distant ones would appear fainter. This was the assumption Herschel made in determining the shape and size of the Milky Way. In truth, objects in the heavens differ in their luminous output as well as in their distance, and it is usually a tricky business to determine whether a faint object is just a powerful star at a great distance or a weak one nearby.

But if we know the distance to a star, by measuring its annual parallax for instance, we can use the distance and the observed brightness of the star to evaluate how luminous it really is. Our nearest star, the sun, whose distance is about 93 million miles, emits an enormous amount of energy each second, enough to keep a hundred million average light bulbs (the 100-watt kind) lit for a billion years. The more distant stars in the nighttime sky, by comparison, turn out to have a wide range of luminosities. A small minority are prodigal suns—ten or a hundred thousand times more luminous than our own. A larger number are equal to the sun in power. And the vast majority are far weaker, producing light at only a few thousandths or even a few ten-thousandths of the solar rate.

How a light varies with time is also a clue to the nature of its source. The lighthouse emits brief bursts of light with long pauses between them; the airplane warning beacons brighten and dim once every few seconds; floating marker buoys keep time to a still different rhythm. Stars vary too. Some flicker up and down in minutes or hours. Others brighten and fade over months or even years. Supernovae are extreme cases, exhibiting a huge increase in brightness in less than a day followed by a slower decline over many months to permanent obscurity.

Later on we will find that the way in which light from a variable star changes (its "light curve") is often a key to understanding it. For instance, pulsars are stars that produce regular sharp pulses of light much like lighthouses. The similarity is instructive; astronomers believe that pulsars are actually spinning stars that emit strong narrow beams of radiation, sweeping their light around the sky just as a beacon sweeps the coastal waters.

Color is also an indication of what the source is like. But at first glance the stars do not seem especially distinctive in this regard. A few well-known stars, Aldebaran in Taurus, Antares in Scorpio, Betelgeuse in Orion, have a reddish tint, but most stars seem a frosty white to the naked eye. Only under the scrutiny of specialized equipment, like the spectrograph, does starlight reveal a wealth of additional information.

It was Isaac Newton, three hundred years ago, who first noted that white light could be broken into a spectrum with the aid of a glass prism. What we see as white, he concluded, is really a mixture of the various colors of the rainbow. But on closer inspection, as Fraunhofer discovered a century and a half later, some colors are missing. The dark lines Fraunhofer saw in the solar spectrum were to prove as distinctive as fingerprints in the analysis of stars. When Bunsen and Kirchhoff later recognized the absorption patterns of familiar chemical elements such as sodium and calcium in the spectrum of the sun, spectral analysis was on its way to becoming an essential tool of astrophysics.

In order to appreciate the power of this technique, consider what we know about the nature and origin of spectra. Light can be regarded as a wave, spreading outward from a source like the expanding circles we see when we drop a pebble in a still pond. What is "waving," in the case of light, is not a material substance like water, however, but something called an electromagnetic field. An electromagnetic field can cause electrically charged particles, such as the electrons in an atom, to move back and forth, just as a cork bobs up and down with the passage of a wave in a pond. Technically speaking, light is a type of electromagnetic wave, or electromagnetic radiation.

Like water waves, electromagnetic waves can be characterized by the distance between their crests or troughs. We call this the wavelength of the light. The wavelengths of visible light are very small, only a few ten-thousandths of a centimeter or so, and astronomers usually adopt a more comfortable set of units named after the Swedish physicist A.J. Ångström, to express them. One angstrom unit (abbreviated Å) is defined as 0.00000001 centimeter, a distance about equal to the size of the individual atoms that make up the paper in this book. Light at the violet end of the visible spectrum has a wavelength of about 4000 Å; red light has a wavelength of about 6500 Å.

We cannot see wavelengths shorter than violet or longer than red, but there are broad ranges of electromagnetic radiation that lie beyond both extremes. These other forms of electromagnetic radiation may go by different names—gamma rays, X rays, ultraviolet, infrared, or radio waves—but they behave the same as light and they carry information that may not be accessible at visible wavelengths. Ultraviolet light, X rays, and gamma rays have shorter wavelengths than visible light. Radio waves and infrared radiation have longer wavelengths. Astronomers studied Supernova 1987A by observing radiation at all these wavelengths.

Luminous objects, from candles to supernovae, emit a mixture of many wavelengths of light (including wavelengths outside the visible, though we shall limit our discussion to visible light for the time being). When this light is passed through a prism, however, different wavelengths of light are bent by different amounts and the light is separated, or dispersed, into a colored spectrum. The exact distance of a color from one end of the spectrum thus depends on its wavelength. Rather than print spectra in color, working scientists sometimes present them in black and white, usually with blue to the left and red to the right. It is becoming even more common to present a spectrum simply as a graph showing the intensity of light at different wavelengths.

To study the mixture of wavelengths in starlight, astronomers attach a device called a spectrograph to a telescope. The light collected by the telescope passes into the spectrograph where the vari-

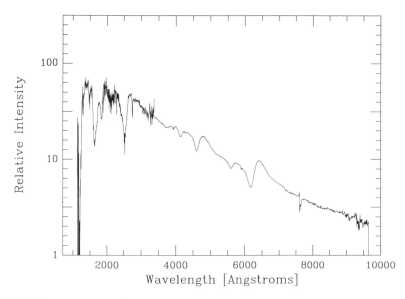

FIGURE 4: A spectrum of Supernova 1987A obtained on February 24, 1987, spanning ultraviolet, visible, and infrared wavelengths. The overall continuous distribution of light is broken by absorption lines (dips) and emission lines (peaks). The deep absorption line near 2600 angstroms is caused by singly ionized magnesium (magnesium atoms missing one electron), and the broad emission line near 6500 angstroms comes from hydrogen atoms. (Courtesy of Ronald G. Eastman, Harvard–Smithsonian Center for Astrophysics.)

ous wavelengths are spread out for study. The spectrum of an ordinary star, for instance, appears as a long strip of light crossed by narrow dark lines. The spectrum of a supernova just after it goes off looks like a long strip crossed by broad bright lines as well as dark lines. (On the spectrum of SN 1987A reproduced here as a graph, the bright lines look like peaks on the graph, and the dark lines look like dips.) The differences between the spectra of stars and supernovae tell us important things about what stars and supernovae are like.

The basic idea is rather straightforward. Gustav Kirchhoff, in the mid-1800s, summarized his observations of laboratory spectra in a series of laws that enabled one to understand the overall physical nature of an object just by looking at its spectrum. Hot solid bodies and gases under high pressure, he noted, produce light at every wavelength. Their spectra appear continuous—the spectrum from the hot filament of an incandescent bulb is a good example.

If a container of cool gas is placed between the light source and the spectrograph, however, dark lines appear across the continuous spectrum. Atoms in the gas absorb light from the continuous source behind them, but only at specific wavelengths determined by the structure of the atom. This produces a distinctive pattern of dark lines, an absorption spectrum, which is characteristic of the type of gas that is doing the absorbing. Hydrogen atoms, for instance, absorb light at wavelengths of 6563, 4861, and 4340 Å, whereas sodium atoms absorb light at 5890 and 5896 Å. A trained spectroscopist can look at a spectrum and give a quick assay of what major gases are present. The pattern of lines is like a fingerprint.

Finally, if the absorbing gas in the container is heated so that it is actually hotter than the source of light behind it, or if the gas is simply viewed against a dark background, a pattern of bright lines is seen. The atoms in the gas are emitting light at certain specific wavelengths, and so a bright-line spectrum is often called an emission spectrum.

Applied to objects in the sky, Kirchhoff's laws enable us to probe their makeup without ever seeing them close up. Stellar spectra, we have noted, are continuous strips of light crossed by dark lines. We can conclude that stars are dense balls of incandescent gas (which produces the continuous spectrum) surrounded by a layer of cooler gas (which produces the absorption lines). We can also match the wavelengths of the lines to those produced by common chemical elements in laboratory experiments: lines of sodium, calcium, and hydrogen are among the most noticeable. The cooler, absorbing gas layers in the outer regions of a star may be thought of as a sort of atmosphere.

Supernovae show a series of bright lines (as well as dark lines) in their spectrum. The bright lines come from very hot clouds of gas, the debris blown away from an exploding star. Other features of the spectrum, such as the notable breadths of the lines, are indications of the extreme temperature and turbulence in the gas. In fact, as we shall shortly see, we can use spectra to measure the speed of the ejected material.

The development of spectral analysis enabled astronomers to make distinctions between objects that appear in the classic catalogs of nebulae compiled by Messier and Herschel. Some of the nebulous or hazy objects, like the Crab Nebula in the constellation of Taurus and the great glowing cloud in the sword of Orion, showed a spectrum consisting entirely of bright lines. William Huggins, a pioneer spectroscopist who first observed the nebulae, reported in 1864 that they were obviously clouds of thin gas, glowing bright because of some internal source of energy. In many cases, we know today, gaseous nebulae are energized by the presence of hot young stars within them. Often these are near regions where new stars are presently being formed from the interstellar gas. But some nebulae, the Crab notable among them, are formed of the debris of stellar explosions— they are the remnants of supernovae.

Finally, some of the objects in the catalogs, especially the impressive clouds of light called spiral nebulae, showed continuous spectra crossed by a few absorption lines. These, astronomers correctly concluded, were large clouds of stars, and the spectra we saw were due to the combined light of millions or billions of suns. In the 19th century, however, no one knew for sure how far away the spiral nebulae were. Were they part of the Milky Way? Or were they distant milky ways, separate from our own local cloud? We shall return to this question soon, but let us first continue to see what else we can learn from spectra.

Using spectra we can take a star's temperature from afar. The continuous spectrum produced by luminous bodies does not contain equal amounts of all wavelengths. A cooler body emits more light near the long-wavelength, or red, end of the spectrum, whereas a hot body emits more light near the short-wavelength, or blue, end of the spectrum, and appears white to our eyes. At low temperatures, in fact, much of the radiation from a source is at infrared wavelengths, and at very high temperatures, at ultraviolet and even X-ray wavelengths. Thus, the distribution of energy in a spectrum gives us information on the temperature of the source. A very cool star should appear somewhat reddish to our eyes, a very hot star, bluish white. This is a bit of knowledge that predates spectral analysis: every blacksmith and steelworker is aware that "red hot" is cooler than "white hot. "

Our sun is somewhere in between; it emits its peak radiation intensity in the yellow part of the spectrum, at a wavelength of about 5500 Å, corresponding to a surface temperature of about 6000 K. (K represents Kelvin, or absolute degrees. For the high temperatures encountered in stars, which exceed those found in blast furnaces, Celsius and Kelvin temperatures are essentially identical.) There are stars that emit most of their radiation shortward of the visible, in the ultraviolet. These are the hottest stars; they appear blue-white to the eye, and have surface temperatures as high as 50,000 K. The coolest stars emit copious amounts of infrared radiation; they appear as reddish stars, and have surface temperatures of 2000 to 3000 K.

If the temperature of a star is known and the energy output (its *luminosity*) can be estimated, it is also possible to estimate its size. You might think that this could simply be done by looking at it through a large telescope under high magnification. Not so: the stars are simply too far away—even through the largest telescopes, all stars look like unresolvable points. But it is possible to use the laws of radiation to determine how big a star must be.

The cooler a star is, the less energy it emits from each square centimeter of surface. Thus, compared to a hot star, a cool star must be a very large object if its total luminosity is to be high. Some stars must be enormous by this measure. Betelgeuse, in the constellation of Orion, is an example of a reddish, cooler star. Though it is rather distant (about 500 light-years), it is one of the brightest objects in the nighttime sky. Betelgeuse must therefore be intrinsically very luminous, despite its low temperature. Astronomers calculate that its diameter is nearly a thousand times that of the sun. If it were placed at the center of the solar system, the Earth's orbit would lie far below its surface. Betelgeuse is an example of a type of star called a red supergiant. These stars, astronomers believe, are nearing the end of their lifetimes, and may die in a supernova blast.

Other stars are, on the other hand, very small objects. There are stars known as white dwarfs that are roughly the size of Earth. The very smallest stars are about the size of Manhattan Island. Called neutron stars, they are thought to be the end products of certain types of supernovae.

In the late 1920s and early 1930s astronomers began to develop intricate methods of analyzing the precise chemical composition of stars and nebulae from the patterns of the chemical absorption lines in their spectra. The first results were a bit of a surprise. Hydrogen, the simplest chemical element with just a single proton in its nucleus and a single electron in orbit, was by far the most abundant element throughout the stars. On Earth it was relatively rare, but in the stars nine of every ten atoms were hydrogen atoms. The next most abundant element was helium, and the remaining chemical elements—carbon, oxygen, iron, aluminum, and so on—were present only in trace amounts.

These proportions seem to be universal; with a very few exceptions, they vary only slightly from star to star. Thus, a typical star can be described, with little exaggeration, as just a hot ball of hydrogen gas, distinguished from its neighbors mainly by just how hot, how large, and how densely packed together it is.

August Comte, who expressed such skepticism about ever learning the physical makeup of the stars, would no doubt have shaken his head in amazement at the last century and a half of progress. Spectral analysis now routinely yields evidence that the remote stars are chemically and physically similar to the sun. Knowing what they are in general, however, is only the beginning of the story. Explaining the differences between stars and understanding the processes that form them, make them shine, and finally end their lives (sometimes as supernovae), remains a major enterprise of astronomy.

THE DOPPLER SHIFT AND THE MOTIONS OF STARS

One of the most fruitful techniques for the analysis of stellar spectra was suggested to the members of the Royal Bohemian Scientific Society on May 25, 1842. The lecturer was Christian Doppler, professor of mathematics at the University of Prague; his subject: "Concerning the colored light of double stars and of some other heavenly bodies. " Doppler argued that if a source were to emit light at a particular wavelength, an observer receiving the light might measure a rather different wavelength. The difference between wavelength emitted and wavelength received would depend on the relative motion of observer and source. If the source—a star, for instance—were moving away from the observer, the received wavelength would be longer than the emitted wavelength. If, on the other hand, the star were approaching the observer, the received wavelength should be shorter. In principle, said the professor, if we knew the amount of change, we could calculate the speed at which the star was moving. The same phenomenon should be noticeable, said Doppler, for wave motion of any sort, even for sound waves, where the shift of wavelength should make itself evident as a change of pitch.

There was quite a bit of initial skepticism among Doppler's contemporaries. To test the theory, the Dutch meteorologist Christophorus Henrickus Didericus Buys Ballot mounted a dramatic experiment. He assembled a brass ensemble on a moving railroad flatcar. Along the roadbed stood a jury of listeners selected for their ability to discriminate absolute pitch. Did the pitch of the music change as the train approached, passed, and receded from the audience? The experts agreed that it did. Thus, with appropriate fanfare, a powerful new technique became available to astronomers.

The change in wavelength that Doppler predicted, simply called the Doppler shift, could be measured by comparing the wavelengths of the spectral lines in the light from a distant star with spectral lines from a stationary lamp mounted on the telescope. In practice the shifts were very small. Consider, for example, a star moving away from our sun at a speed of 30 kilometers per second, roughly 100,000 miles per hour. This may seem extremely rapid, but it is a rather typical value for stars in the Milky Way. The dark lines from sodium in such a star should be shifted no more than $1/2$ Å from their normal wavelengths of 5890 and 5896 Å (a change of a hundredth of one percent) by Doppler's effect.

Doppler himself was aware of the minuscule shift to be expected from a star moving at 30 kilometers per second. But he believed that stars often moved far more rapidly than that, perhaps even several thousand times faster. If this were the case, the shift in wavelength would be so large that it would actually change the overall color of a star. This, he proposed, is why some stars appear to be different colors. A star approaching us fast enough would appear to be violet and a star receding would be red. In reality, the Doppler shifts for stars are far too small to show this effect; the color differences he sought to explain actually arise from differences in temperature. Yet Doppler had no way of knowing this; he proposed his theory, but made no observations of stellar spectra himself. It was just as well. The shifts of individual spectral lines were, at the time, almost impossible to measure. But there have been dramatic improvements in measuring devices since then, and today the Doppler shift is one of the most powerful tools of astronomy.

A quarter of a century passed between Doppler's Bohemian lecture and the application of his ideas to stellar spectra. The English astronomer William Huggins, in 1868, announced that he had measured the velocity of the bright star Sirius by comparing the wavelength of a stellar hydrogen line to the spectrum produced by a sample of hot hydrogen gas. Sirius, Huggins claimed, was speeding away at 47 kilometers per second. Just how difficult this measurement must have been is indicated by modern figures: Sirius is actually approaching us at about 8 kilometers per second. Huggins's visual inspection of its spectrum just wasn't precise enough to get a good result.

Yet had early observers like Huggins not persisted in their efforts to observe and refine Doppler's technique, our knowledge of the motions of the stars would have advanced more slowly. Huggins observed spectra without the aid of photography. It soon became common among astronomers to record spectra photographically and to measure the wavelengths of spectral lines in the photographs using precision microscopes mounted on carefully machined screws. Lately, astronomers have made the process even more convenient by capturing spectra on computer-compatible TV detectors and measuring the Doppler shifts directly in a digital computer. Measurements of stellar velocities, as a result, have become almost routine. (Doppler measurements of a different sort also have become routine along American highways. The "radar" guns wielded by state troopers send out a radio signal of known wavelength and use the shift in the returned signal bounced off a moving vehicle as a measure of its speed.)

The Doppler shift finds many uses in the analysis of light from supernovae. Gases expelled from an exploding star travel at enormous speeds, and their Doppler shifts are easily determined. In addition, by studying the Doppler shift in ordinary stars, we have learned much about how ordinary stars behave, and under what conditions they are likely to explode.

In Chapter 1 we noted that the stars move with respect to one another, showing a gradual motion (called their proper motion) that becomes apparent when star positions measured many years apart are compared. The Doppler shift in stellar spectra, however, yields the speed of the star immediately, without requiring a wait of decades or centuries. There is a distinction, however. The Doppler shift measures only the speed of approach or recession of a star; the proper motion measures only the motion of the star perpendicular to our line of sight. Usually a star does both, and astronomers must combine proper motion measurements with Doppler shift measurements to determine the true motion of the star through space.

Doppler shifts have been used to demonstrate that many stars that appear to be single actually have companions. Pairs of stars that appear close together are not difficult to spot. A well-known example is Mizar, the middle star in the handle of the Big Dipper, which has a close companion star that can just be distinguished by the naked eye on a clear night. But how can we be certain that such neighboring stars form a true physical association, locked together by gravitation as the planets are locked to the sun? Couldn't we be seeing a chance alignment of two stars, one of which was far closer to us than the other?

In some cases, only time was required to answer the question. Between 1784 and 1804, William Herschel, among the most patient of observers, recorded the relative positions of about 50 close pairs as they slowly revolved around each other. That painstaking technique, still used by a few dedicated researchers, works only for some of the nearest binary stars. If a pair of stars is too far away from us, or if the individual stars are too close to one another, they blend into a single point and are indistinguishable from a solitary star.

Indistinguishable, that is, until looked at with a spectrograph. Then we can detect changing Doppler shifts from binary stars. It comes about this way: imagine yourself at the races, sitting at one end of a large oval track. A horse barrels down the straightaway toward you, rounds the turn across your line of sight, and gallops away toward the far turn. In similar fashion, each member of a binary system moves alternately toward and away from the observer as it orbits around its companion. (I have assumed here a peculiar oval shape for the orbit and a particular orientation for the observer. Though somewhat contrived, this situation is easy to visualize and illustrates the principle involved.) Even if individual stars cannot be separated in a telescope, the combined light from a pair carries a message of binarity. One can see a regular shift in the wavelength of spectral lines as they are alternately shifted toward the red and the blue end of the spectrum by the changing velocities of each member of the pair.

Thousands of covert binary stars have been recognized as a result of regular variations in their Doppler velocities. It's estimated, in fact, that as many as half of all stars, including those that become super-novae, may be binary. In a number of cases, in fact, we have discov-ered neutron stars—which are produced by supernovae—in binary systems. Some supernovae, called Type I explosions, may actually require a binary star. These supernovae, according to the most widely accepted theory, occur when one member of a binary pair deposits explosive nuclear fuel on the surface of the other. (We shall present the details in Chapter 6.) Though no one has yet observed a Type I supernova in a previously known binary system, Doppler shifts will eventually provide the crucial evidence to test the theory.

THE MEASURE OF THE UNIVERSE

Brightness, spectrum, variability. Vital statistics of the stars are written in the language of starlight. The same language also tells us the scale of distances in the universe. Though the subject of cosmic dimensions will be the focus of a later chapter, we'll be able to appre-ciate the intervening narrative if we begin with some idea of where we're headed. The early chapters of the supernova story are set among the stars of our home, the Milky Way Galaxy. The later chap-ters take place on an immeasurably larger stage: a universe of other galaxies stretching to the limits of our vision, billions of light-years away.

We have referred already to the Milky Way, the name given to a feeble luminous ribbon of light that encircles the nighttime sky. If you've lived all your life under city lights, the Milky Way may be more legend than experience, for it takes a dark, moonless sky to show it to advantage. Every star visible to the naked eye is a member of the Milky Way, including the sun. But the vast majority of fainter stars in this giant assemblage make themselves known mainly by their indistinct glow. Binoculars or a small telescope can reveal its nature: billions of faint, distant stars, most of them less luminous than the sun, massed into a ragged, irregular band.

It's easy to tell that we are in the midst of a cloud of stars, for the Milky Way cuts a swath around the entire celestial sphere. If we were at the edge of the cloud, it would only appear off to one side of the sky. It's obvious, also, that the Milky Way is a rather flat cloud, for if it were spherical in shape, we'd see a glow all over the sky, rather than just a single ribbon of light. But where is its center, and where are its edges?

The first person to address this problem directly was William Herschel, working in England during the years of the American Revolution. We spoke of Herschel's method of counting, or "gauging," the stars in Chapter 1, and of his tentative model of a disk-shaped galaxy. This "grindstone" concept of the Milky Way, the result of years at the telescope, was a remarkable achievement, especially in comparison with the astronomy of 250 years earlier. In 1543 a Polish cleric, Nicolaus Copernicus, had proposed a universe only a bit larger than our solar system, bounded by a sphere of fixed stars. Not surprisingly, the Milky Way was unimportant in Copernicus's universe, as well as in the cosmologies that preceded it, for the stellar nature of the Milky Way was unknown until Galileo first observed it with a telescope in 1609.

Herschel's model, on the other hand, was firmly grounded in nightly telescopic observations. He and his sister had gauged the universe, actually counted its stars, finding not a single sun, but an immense system of millions. In his later years, Herschel recognized the shortcomings of this scheme. He acknowledged that his telescopes had not penetrated to the very edges of our star system. Yet even the limited model he had ventured was vast beyond ordinary powers of comprehension.

A century after Herschel's death, at the dawn of the 20th century, the pace of astronomical development had quickened. Photography, spectroscopy, and the use of large telescopes were well-established techniques. Numerous parallaxes for stars had been measured; the spectrum of the sun had been studied in detail; some of the fundamental characteristics of stars and nebulae were well understood. The time was ripe for a revision of Herschel's work, for a better gauging of the universe.

Measuring distances remained a problem. For stars beyond a few hundred light-years, the annual parallax shift degenerates from a sway to a twitch, impossible to measure even with the most meticulous scrutiny. The vast majority of stars lie beyond such scrutiny, inaccessible to the yardstick of parallax. But the stars within a few hundred light-years, whose distances and luminosities *were* well known, provided a means of reaching out to greater distances.

Imagine taking one of these stars of known luminosity and nudging it farther and farther from your telescope. At larger distances, the star appears fainter, and you can tell how far the star has receded from its original position by just how much fainter it is. A star moved to twice its present distance, for instance, dims by a factor of four; a star moved to three times its distance dims by nine times, and so on. (Because the brightness decreases with the square of the distance, we call this the "inverse square" law for radiation.) So if you know how much light a star emits, the amount you receive is a measure of its distance. The star acts, astronomers are fond of saying, as a "standard candle."

But how can you tell how luminous a star really is? The answer lies, in many cases, in the spectrum of a star. Stellar spectra show characteristic patterns of lines. By the first decades of this century astronomers were beginning to recognize that the patterns were related to the inherent luminosity of the star. An intrinsically powerful star could be recognized by the presence of one particular set of dark lines (and also by how sharp and well defined the lines appeared), a weak star by the presence of a different array of lines. Once the "spectral type" was known, one had a pretty good idea how much light the star was really emitting, and thus how far away it was.

This method of gauging the stars harks back to Herschel's original method of simply counting the stars of a given brightness. But whereas Herschel assumed that all stars gave off the same amount of light, the acknowledgment of individual spectral types takes into account the fact that some stars are more powerful than others. Developed through the decades, it enables us to map out the distances to individual stars and clusters of stars in the Milky Way, out to distances of several tens of thousands of light-years.

Even with this method of measuring distances, however, astronomers in the early years of this century had a myopic view of the system of stars in which we live. For a century after William Herschel, most astronomers would have put Earth near the center of the grindstone, and the edges several thousand light-years out.

It fell to Harvard astronomer Harlow Shapley, a young and spirited midwesterner, to revise these notions. Shapley's research involved measuring the distances to some curious objects that had appeared in Messier's catalog: globular clusters, giant systems of several hundred thousand stars held together by the force of gravity. Seen through a telescope a globular cluster appears as a strikingly circular smudge of luminescence, indistinct at the center, thinning to a scattering of stars at the edges. Oddly, unlike the majority of the stars in the Milky Way, the globular clusters seemed to be concentrated to one side of the sky, roughly in the direction of the constellations Scorpius and Sagittarius.

Why was this? Shapley knew that only by measuring the distances to the globulars could he get a three-dimensional perspective of what was going on. Because the clusters were too far away to measure their annual parallax, Shapley tried a more indirect plan of attack. Occasionally, among the spattering of faint stars near the edges of a globular cluster, one or two might be noted that varied regularly in light, like the pulsing beacon on a radio tower. Shapley proposed using these regular variables as standard candles to measure the distances to the globular clusters.

Shapley recalled the work of another Harvard astronomer, Henrietta S. Leavitt. In 1912, Leavitt had published a study of a type of variable star, the so-called Cepheid variables, in the Large and Small Magellanic Clouds, two nearby irregular swirls of stars that can be seen in the southern sky. Cepheids brighten and dim regularly over periods of several days to several weeks. They are readily recognizable by their rapid rise to peak brightness and their rather gradual decline. Leavitt was struck by the relation between the length of the periods of Cepheids in the Magellanic Clouds and their apparent brightness. The Cepheids of short period seemed to be fainter than those of long period. Simply by timing the period of variation of a Cepheid, suggested Leavitt, one could determine how luminous it was, and thus determine its distance.

Shapley calculated the periods and luminosities of relatively nearby Cepheid variables, then assumed that cluster variables of comparable period had the same luminosities. He was thus able to estimate the distances to 86 individual globular clusters and to investigate their distribution in three dimensions. The globular clusters, he found, were scattered throughout a large spherical volume with its center in the direction of the constellation of Sagittarius.

The center of the globular cluster system, Shapley concluded, was the true center of the Milky Way Galaxy. As he visualized it in his early papers, our galaxy was a thin disk of stars over 60,000 light-years in diameter, surrounded by a spherical halo of globular clusters. The sun no longer held a central position; it was located in the disk about 20,000 light-years from the center. Since Shapley's original work, the dimensions and finer details of the Milky Way have been revised a bit, but the essential features remain the same. Current estimates are that the diameter of the Milky Way disk is about 100,000 light-years, and its thickness, only about 1000 light-years. Imagine two long-playing records stacked together, and you have some idea of the proportions.

Herschel had failed to discover the true shape of the Milky Way in large part because he was unaware that there were clouds of dust and gas concentrated in the disk of the galaxy. Because the sun is located in the disk, we view the Milky Way through a fog. We can only see to a limited distance in any direction along the disk, far less than its true extent, and so we are led to the conclusion that we are in its center. By studying the globular clusters that are located, by and large, far outside the dust-laden disk, Shapley was able to map its true dimensions.

What lay beyond? From time to time there had been speculation that some of the hazy objects or nebulae that had been recorded by Messier and others were giant groups of stars like the Milky Way. Herschel himself, in writing about his measurements of our own galaxy, had suggested as much:

> As we are used to call the appearance of the heavens, where it is surrounded with a bright zone, the Milky Way, it may not be amiss to point out some other very remarkable Nebulae which cannot well be less, but are probably much larger than our own system; and, being also extended, the inhabitants of the planets that attend the stars which compose them must likewise perceive the same phaenomena. For which reason they may also be called milky ways by way of distinction.

In 1845, William Parsons, the third earl of Rosse, constructed a telescope with a 72-inch mirror in the garden of his manor home, Birr Castle, in central Ireland, and used it extensively to observe the nebulae. Lord Rosse's instrument, unwieldy and dangerous to use (the observer had to balance on a platform high above the ground while servants pointed the telescope by pulling on ropes), surpassed all the telescopes of its day in its ability to collect starlight. Thus, he was able to distinguish details in the hazy nebulae, and to recognize that they displayed a wide variety of shapes.

Notable among the objects Rosse observed were the spiral

nebulae, which appear in his early sketches as pinwheels of light, trailing out sprays of luminescence like Roman candles at a fireworks display. The spectra of spirals, later observers discovered, resembled the spectra of large collections of stars. Looking at them through telescopes of high magnification, however, failed to separate their luminescence into individual points of light. If they were made of stars, they had to be very distant.

The question was, how distant? If the spiral nebulae were only a few thousand or a few tens of thousands of light-years from us, then they were inhabitants of the Milky Way, on a par with the globular clusters. If they were farther away, then they might be as large as or larger than our own system.

Shapley initially favored the notion that the spirals were relatively nearby objects. In April 1920 he shared the platform with an astronomer from Lick Observatory in California, Heber D. Curtis, at a debate sponsored by the National Academy of Sciences, in Washington, D.C. Shapley argued that the globular clusters marked the

FIGURE 5: NGC 6946, a spiral galaxy in the constellation of Cepheus. (National Optical Astronomy Observatories.)

outer boundary of the entire stellar system. Spiral nebulae, he claimed, were members of the Milky Way. Curtis argued that they were extragalactic, located at least several hundred thousands of light-years from us, and that if the Milky Way were looked at from that distance, it would appear like a typical spiral nebula. When the meeting adjourned, there was still no meeting of the minds; the evidence was much too meager to reach a conclusion one way or the other.

Ironically, the study of the light variations of Cepheids, the same technique that Shapley had used to survey the globular clusters, soon provided conclusive evidence for the extragalactic distance scale he had argued against. In 1923, Edwin Hubble, an astronomer at Mount Wilson Observatory, on a mountain above the small California town of Pasadena, managed to resolve the outer regions of two of the brightest spiral nebulae, M31 and M33 (entries 31 and 33 in Messier's list), into individual stars.

Spiral nebulae are difficult to photograph because their light is so spread out, or diffuse. Only if the sky is extremely dark, free of the glare of the city, will the faint light of the nebula contrast strongly with the general background glow of the sky. Today Pasadena at night is ablaze with light, and the large telescopes at Mount Wilson are useless for the kind of work Hubble did. But in his day the site was a fine one, and Mount Wilson's telescopes were state-of-the-art. Its largest reflector, with a mirror 100 inches in diameter, was from the 1920s to the late 1940s the largest telescope in the world.

Hubble made good use of Mount Wilson's large telescopes and transparent skies. He monitored M31, a great spiral in the constellation of Andromeda, on a regular basis. In repeated photographs, some of its faint stars exhibited distinctive variations in brightness—a rapid increase in brightness of several magnitudes followed by a gradual decline—the signature of a Cepheid variable.

Timing these variations enabled Hubble to determine the inherent luminosity of the Cepheids and, from their observed brightness, to estimate the distance to the Andromeda nebula. His initial estimate in 1924, based on observations of just two variables, was a bit under a million light-years. (The currently accepted value is about twice as great, for since then there have been major recalibrations of the relation between the periods of Cepheids and their luminosities.) This placed the nebula far outside the Milky Way. If it was that remote, it was surely a body as large as our own galaxy. The universe took on mind-stretching proportions.

Hubble's later work confirmed his initial impression that spiral nebulae were distant galaxies like the Milky Way. Using Cepheids and other bright objects as standard candles to measure the distances of galaxies, he was able to confirm that a universe of galaxies extended for millions of light-years in all directions. In addition, he began to recognize distant galaxies that were not spiral in shape. Some, which Hubble called elliptical galaxies, looked like large-scale versions of globular clusters, huge spheres or egg-shaped aggregations of billions or even trillions of stars. Other galaxies were irregular in shape. The nearest galaxies were the Large and Small Magellanic Clouds, irregular galaxies only a bit over 100,000 light-years away, just beyond the edges of the Milky Way.

At Lowell Observatory in Flagstaff, Arizona, astronomer Vesto Slipher had also been looking at galaxies. Slipher was interested in their speeds, however, not their distances. By 1925 he had obtained velocities for about 40 galaxies. Strangely enough, with few exceptions, all the Doppler shifts were to longer or redder wavelengths. The conclusion: virtually all the galaxies were receding from us. If the motions of the distant galaxies were truly random, like the jostling of balls in a lottery basket, one would expect to find as many galaxies coming toward us as moving away. The universal redshift was a puzzle.

Then Hubble made a remarkable discovery. Comparing the velocities measured by Slipher with the distances he had determined, he noted a clear and compelling pattern, known today as the Hubble relation. Stated succinctly: the more distant the galaxy, the faster it is moving away from our own.

With the announcement of Hubble's discovery, the significance of Slipher's redshifts became clear. The universe, astronomers realized, was expanding, and the galaxies were being carried apart on the ballooning framework of space itself. According to the generally accepted interpretation of this expansion, called the big bang or "standard model" of the universe, it all began sometime between 10 and 20 billion years ago, when the entire universe exploded from an unimaginably hot and dense state. Since then, everything has been thinning out as the galaxies speed away from each other.

Hubble's relation could then be turned around and used to probe deep into space, out to where galaxies appear as mere smudges of light in the largest telescopes. The faster a galaxy is receding from us, the more remote it is. Take a spectrum of the galaxy, then measure the wavelength shift, and one has a measure of its distance.

At the limits of our telescopic vision, galaxies are as numerous as stars. A hundred billion or more are accessible to the largest telescopes. So the task of measuring galaxy redshifts is comparable to the cataloging and position measuring carried out by 18th and 19th century astronomers. It is a task performed today with large telescopes and computerized spectrographs. It has been well established that galaxies are grouped into clusters, much as stars are grouped into galaxies. Half a century after Hubble began his work, the three-dimensional mapping of the universe is just beginning to reveal its large-scale organization. The clusters of galaxies themselves seem to be part of still larger structures, superclusters, hundreds of millions of light-years in size, whose general form is still uncertain.

FIGURE 6: Part of the Virgo cluster, a large cluster of several thousand galaxies located about 50 million light-years away from our Milky Way Galaxy. (National Optical Astronomy Observatories.)

In all this vastness, one would think that a supernova would shrink to inconspicuousness. What can a single exploding star mean to a universe of galaxies? Yet at the cosmic scale, supernovae seem far more significant than they do from Earth. They are frequent and ubiquitous. That is not, of course, the impression we get by looking at the sky. Within our galaxy we see only the most nearby supernovae. These occur very seldom and are sometimes hidden behind intervening clouds of gas and dust. But seen from a distance, supernovae are more frequent and conspicuous. For a few brief weeks a single supernova may outshine a galaxy of stars. It's estimated that a supernova occurs several times each century in an individual spiral galaxy—that's often in a universe fifteen billion years old. With billions of galaxies visible to a cosmic observer, supernovae would be going off all the time.

It is largely a matter of perspective. If we had the power to view the universe all at once on the largest scale, as a mountain climber surveys a valley below, and if fifteen billion years seemed like a day to us, we could readily appreciate the role of supernovae in the overall scheme of things. Here and there among the multitude of galaxies we would see supernovae flaring up, as numerous as fireflies on a summer lawn, momentarily outshining a hundred billion neighbor suns.

But we study the universe from inside, rather than outside, and our viewpoint is bound to planet Earth, deep within the Milky Way. It is there that we begin our story of supernovae, at a time when the cosmos was still regarded as a rather small place, and when the stars spoke directly to men.

CHAPTER 3

New Stars in Ancient Times

> You'll wait a long, long time for anything
> much
> To happen in heaven . . .
> —Robert Frost, *On Looking Up By Chance at
> the Constellations*

Long before the birth of modern astrophysics, the light of supernovae fell on uncomprehending eyes. Yet there were people who looked up attentively at the heavens back then, people whose writings still have much to teach us. Although in a scientific sense they understood little of what they saw, they were aware of the sky, sensitive to its changing patterns, and curious about its meaning. That is reason enough to value what they wrote, for it awakens us to a common thread of humanity linking present and past.

Remarkably, the ancient records still function as an important resource for modern science. A recognizable description of a supernova provides, at the very least, one precious item of data: the time at which the light from a stellar explosion first reached our planet. If there is additional information on the brightness of the object, its color, and the length of time it remained visible, we may be able to get an idea how far away it was and what kind of star it was before it exploded. And from the number of such reports in the past we can learn how often supernovae occur. Supernovae are dying stars, and historical sightings thus yield estimates of the death rate among stars in our galaxy, a vital statistic as important to astronomy as tables of mortality are to medicine.

What is more, knowing the dates and positions of the historical supernovae allows us to follow the slow progression of effects produced by the supernova blast. For tens of thousands of years after the initial flash of light has faded, a cloud of debris from a supernova explosion dissipates gradually into the surrounding interstellar space. In our region of the Milky Way, well over a hundred such remnants have been observed. Some are old, some are relatively young. Some have spent their energy rapidly as they pushed their way outward through billows of preexisting gas; others have expanded with little resistance into a virtual vacuum.

Supernova remnants are, in short, products both of their history and of their environment. To understand their varied properties, we must have as much information as possible about them. Just as a series of photographs of infants, youths, and adults, properly arranged, can give biologists a knowledge of human growth and maturation, so the remnants of supernovae embody the dynamic history of dying stars—if only we are able to distinguish the old from the young, the commonplace from the pathologic. An ancient astronomical account can supply that information, but first it must be recognizable as describing a supernova explosion, and, equally important, it must contain sufficient detail on the position of the explosion so that we can identify it with a known supernova remnant.

Stuck where we are in space and in time, we have no choice but to rely on the testimony of observers from times long past. We must turn to ancient documents, written in languages long forgotten, filled with the cultural imagery of times gone by. We attempt to extract, from a welter of scientifically extraneous information, accounts of supernovae that were regarded by their beholders merely as "guest stars," or "new stars," or omens of good or ill fortune. To make matters more difficult, star positions in these documents use constellations and systems of measurement no longer familiar to any but a few scholars of antiquities. The words of the ancient chroniclers have become virtually unintelligible to their spiritual descendants, the astronomers of the 20th century.

THE EARLY RECORDS

Modern stargazers see a different universe from that of astronomers of earlier civilizations. I am not speaking of the fact that the stars move, or that the skies are a bit more polluted today than they were two or three thousand years ago. Rather I am speaking of the different frame of mind with which modern people look heavenward. When we read ancient astronomical texts, whether they describe something as familiar as the rising of the moon or as exotic as a supernova, we bring with us modern sensibilities that color our inter- pretations and temper the voices of the scribes. We therefore find it difficult, and sometimes impossible, to extract useful information from the writings of the past.

Consider a few differences: Ours is an age of air travel, of photos of Earth from space, of communications that span the globe in a fraction of a second. We carry the precise time in microchips on our wrist; we consult the evening news for knowledge of present, past, and future events. We look at the universe with an insight born of centuries of scientific research, leading to the current view of a uni- verse of superclusters of galaxies vast beyond ordinary comprehen- sion. Our ancestors, by contrast, lived in a time when the boundaries of the world extended not far beyond the natural barriers of moun- tains and shoreline; when Earth was the center of creation; when history, as recounted by the village bard, was measured in terms like "long ago," when a sequence of precise dates and eras was unknown. To our forebears the sun really rose each morning; to us, it just seems that way.

In those times there was no science as we know it, a discipline separate and distinct from the everyday world of human affairs. Nor, most scholars agree, were our ancestors much aware of an objective reality distinct from the observer, amenable to measurement and ma- nipulation. The earliest accounts of the heavens, which date back to the first written documents in Egypt, Babylonia, and China, don't offer us much hard information about the appearance of new stars in the heavens.

That is not to say that new objects were not mentioned. On the contrary, celestial events were recorded, but regrettably there was little uniformity to the record keeping, little attention to what we would consider essential facts and figures. Pliny's reference to Hip- parchus's new star, for instance, is brief and uninformative. The Greek astronomer, he says (writing, incidentally, two centuries after the fact), "discovered a new star and another one that originated at that time." That is all.

The best known example of this type of reporting is the Star of Bethlehem, an apparition mentioned in only one place in the New Testament: Matthew, Chapter 2. There we read that wise men came to see Herod the King, saying, " Where is he that is born King of the Jews? For we have seen his star in the east, and are come to worship him." Would a modern astronomer countenance such data-taking? Where are the celestial coordinates of the star? Magnitude estimates? Descriptions of form and of color? Later in the same passage Herod "inquired of them diligently what time the star appeared. And he sent them to Bethlehem. . . . When they heard the king, they departed: and lo, the star, which they saw in the east, went before them, till it came and stood over where the young child was."

Such references, maddeningly vague by modern standards, are common in the older texts, and generate more heat than light in scholarly circles. Did the star move against the background of constellations? Or did it "go before them" in the sense that it was visible in the east just after dusk or just prior to dawn? Could it be that the Star of Bethlehem was not a star at all, but a symbolic reference to an astrologically favorable alignment of planets? Or even that it was a rhetorical device to legitimize Jesus in the eyes of skeptical Zoroastrians, who believed in the power of astrology? There is little hard information to go on. In astronomical terms, the Biblical passage by itself is virtually useless. It may have been a comet; it may have been a supernova; it may have been an astrological sign intelligible only to the Wise Men from the East. With so little explicit evidence, I doubt that we shall ever know for sure.

We should not, however, fault our ancestors for their imprecision, for science was not the purpose of their observation. The heavens, to most ancient cultures, were a celestial billboard on which divine messages were displayed. Symbolism and fact were blended inseparably in every encounter with the natural world; thus, it was senseless to devote much attention to such superficial aspects of an event as its brightness, position, or time of appearance. Light, space, and time were merely the medium for the message writ in starlight. Accordingly, Matthew and those who accepted his gospel regarded the *meaning* of the "star"—the announcement of the coming of the Messiah—and not its brightness or position, as the real news for humankind.

Even if we could learn to think like ancient astronomers, we would still be separated from the past by the ravages of time. Many of the old records survive not as single documents, but as fragments, dispersed to museums, and largely unread. Even where a large quantity of material exists, much remains to be transcribed and translated.

This is particularly true of the records of the Babylonians, who were, for a long period, assiduous watchers of the skies. Their cuneiform records, incised into tablets of clay, lie broken and scattered in Oriental collections in various countries. Future scholars may uncover a wealth of useful information in the unstudied texts of the Middle East. But for the present, the most valuable source of information on the ancient state of the heavens comes from one source: the historical records of the Chinese court, which provide us roughly two thousand years of astronomical observations. Important records from somewhat later times are also preserved in Japanese and Korean historical literature. Though we pride ourselves on our technical sophistication, Western records yield little of comparable interest until approximately 400 years ago.

THE CHINESE RECORDS

In China, from about the 2nd century B.C. onward to the current century, there flourished a tradition of astronomical record-keeping that has proven to be remarkably complete and exceedingly informative. Rooted in the myth and mysticism of the past, it continues to nourish modern-day research into the nature of supernovae, comets, and other celestial events long ignored or forgotten by Western observers. Though modern scholarship has belatedly come to realize the enormous debt Westerners owe to the Chinese in a wide range of technical areas (they were the first to develop printing, steel making, and deep well drilling, to name just a few), we are still surprised that they developed a set of extensive astronomical records when other cultures did not.

The Chinese watched the heavens carefully because they believed humankind played an indispensible role in the smooth functioning of the universe. As they saw it, a sympathetic harmony existed between human affairs and events in the sky; human society and the heavens were seen as interconnected elements of a complex, universal organism. (The 19th century British poet Francis Thompson expressed something like this in the line "Thou canst not stir a flower without troubling a star.") One looked at the sky, then, not simply to foretell the future, as some folks consult newspaper astrology columns, but rather to learn what one was expected to do to make the world go smoothly. If one channeled one's life along a cosmically preordained course, the result might not be pleasant, but it would certainly be satisfactory. On the other hand, to go against the signs in the sky would be to violate the prescribed pattern of nature.

Over time, this general philosophical outlook became associated particularly with the personage of the emperor, who as a semidivine representative of heaven, bore the responsibility of guiding the affairs of state along the enlightened path. A passage by the astronomer Shih-sen (4th century B.C.) typifies the Chinese world view:

> When a wise prince occupies the throne, the moon follows the right way. When the prince is not wise and the ministers exercise power, the moon loses its way. When the high officials let their interest prevail over public interest, the moon goes astray toward north or south. When the moon is rash, it is because the prince is slow in punishing; when the moon is slow, it is because the prince is rash in punishing.

The emperor could not attend to the heavens all by himself, of course; like Herod, he relied on astronomical wise men to monitor the heavens and alert him to the latest astropolitical dispatches. In a real sense, the emperor's astronomical ministers functioned like advisors in a modern president's cabinet, advising him on how best to conduct the affairs of state in a harmonious and effective manner.

Astronomy was such an integral part of the state that it merited a special branch of government: an Astronomical Bureau, headed by an imperial astronomer who reported directly to the emperor. For more than 2000 years the Astronomical Bureau operated at least one, and sometimes two observatories in the capital city to monitor the heavens and apprise the emperor of the celestial signs. The failure of an astronomical official to report a celestial event was a serious crime.

As was the case with many ancient cultures, the power of the Chinese court was drawn in part from its maintaining of the calendar, supplying practical information for the timing of planting, harvesting, and practicing religious ritual. This was indeed a major function of the Astronomical Bureau. According to Chinese texts, the legendary Emperor Yao established government-supported astronomy by commanding the brothers Hsi and Ho "in reverent accordance with the august Heavens, to compute and delineate the sun, moon, and stars, and the celestial markers, and so deliver respectfully the seasons to be observed by the people." Though some details of the story are more mythical than historical, it illustrates that calendar making was, from the very beginning, a primary task of the imperial astronomers.

The astrological function of the Astronomical Bureau was just as important. After recording and interpreting the signs of the heavens, the court astronomers would tell the emperor when it was propitious to make war, levy taxes, or embark on a journey. They thus enjoyed a special, intimate relationship with the state. So important was this function that in the year 840, an edict was issued mandating that the imperial observatory should keep all its proceedings secret, even from members of other government departments: "From now onwards, therefore, the astronomical officials are on no account to mix with civil servants and common people in general."

Combining the diligence and secrecy of early-warning radar operators, the Chinese imperial astronomers watched and recorded conspicuous events in the heavens, from comets to supernovae. If only we had their original logbooks (if such existed), for though the level of training and activity at the imperial observatory fluctuated from dynasty to dynasty, there must have been few brilliant events that escaped their notice. All that remain, however, are secondary accounts muddied by the distortions of time and the transcription of later scribes. Fragmentary records date back to 1500 B.C. but relatively regular chronicles did not begin until after about 200 B.C., when Emperor Ch'in Shih-huang ordered the burning of all books and records of all previous dynasties as part of his effort to unify the kingdom and consolidate his power.

Though the records are relatively continuous thereafter, the observations of the imperial astronomers are seldom presented in an impersonal, scientific form. The astronomical reports appear primarily in extensive summaries of celestial events that were a standard feature of the official dynastic histories of the Chinese court, and, moreover, were often assembled several centuries after the fact. These were never dispassionate recitals of dates and positions in the sky; they read rather like imaginative stories in which men, stars, and planets played equal roles. Because heavenly signs were believed to be inseparably associated with earthly affairs, the astronomical observations would often be cited to explain a military victory, a popular revolt, or the sudden death of a monarch, much as a political or sociological analysis would be used in a modern history.

A typical account, from the official history of the Later Han dynasty (23–220 A.D.) reads:

> Eighth year, 10th month, last day of the month [December 16, 65]. There was an eclipse of the sun and it was total. It was 11 degrees in Nan-tou

[the name of a "lunar mansion," a region of the sky that functioned in
some ways like a sign of the zodiac]. Nan-tou represents the state of Wu.
Kuang-ling, as far as the constellations are concerned, belongs to Wu.
Two years later Ching, King of Kuang-ling, was accused of plotting re-
bellion and committed suicide.

Reading the dynastic histories with modern sensibilities, there is
no doubt that the ancient writers were being a trifle selective in their
interpretation. Given the variety of events that occur in the heavens,
and the ceaseless fluctuation of human fortunes, the chroniclers
doubtless had no trouble finding an appropriate celestial event for
anything that happened on Earth and, conversely, a significant victory or an awful calamity for every heavenly sign. Effect seldom followed immediately after cause; years could pass between an ominous
arrangement of the planets and the death of an emperor. And frequently the details of the reports were embellished to cast events in a
harmonious context. For instance, a new star that appeared in 1054
was reported in one history as "yellow and favorable to the Emperor." Was the star really yellow? (There are other accounts that it
was "reddish white.") Coincidentally, perhaps, yellow was the color
generally associated with the Sung dynasty (960–1279 A.D.) emperors, just as red was the color of the Han monarchs. Had the same
star appeared during the Han dynasty a thousand years earlier, one
suspects, it might have been reported as brilliant red.

Nor were the records scrupulously complete. Some celestial
events were more likely to be recorded than others. We have good
reason to believe that when times were hard, imperial astronomers
watched the heavens more carefully than when the going was easy.
In the accounts of the Chin dynasty (265–420 A.D.), for instance,
scholars have found that astrological omens were recorded more frequently during the last decades, when public discontent with the
state was rising, than during its early years. As is the case with
American presidents, the head of state appears to have enjoyed a
"honeymoon" with both populace and cosmos at the beginning of his
rule. Only as state power waned did the emperor turn to the heavens
more and more for divine guidance.

Yet the Chinese records, if neither absolutely complete nor
rigidly impartial, represent a remarkable collection of systematic observations. Long before Hipparchus's star catalog, Chinese astronomers had developed a system of locating stars that produced accurate and relatively consistent results, and that remains intelligible to

the present day. Shih-sen, an astronomer of the 4th century B.C., is said to have produced a listing of 809 stars assigned to 122 asterisms (close, distinctive groupings of a few stars, smaller, as a rule, than constellations). Several hundred of these stellar groupings were eventually employed by Chinese observers, in contrast to the 88 considerably larger constellations officially recognized by Western astronomy today. Surviving Oriental globes and sky maps can be used to identify many of the stars in the Chinese asterisms with their 20th century counterparts, though the common Chinese practice of rendering all stars, bright and faint alike, with identical dots makes the task tedious and sometimes uncertain.

Seeing shapes in the bright stars is something of a Rorschach test; no two cultures connect the dots in quite the same fashion. The old reliable Big Dipper, for instance, has long been called a "plough" in England and a "wagon" in Germany. Early Chinese astronomy, which developed largely in isolation from the West, adopted quite a different, but no less complete, set of conventions for mapping the sky. Where we see creatures from Greek and Roman mythology in the heavens, they tended to see the sky as a panorama of figures from the imperial court. Asterisms included the Emperor, Crown Prince, Celestial Dog, and the Court Eunuchs. In addition to these small groupings of stars, the sky was divided into 28 larger divisions, the so-called "lunar mansions," which functioned somewhat like the signs of the zodiac in Western astronomy. The 12 zodiacal signs (named Aries, Taurus, Gemini, and so on after the prominent constellations within them) are divisions of the sky based on the monthly progression of the sun through the heavens; similarly the 28 lunar mansions, which divided the celestial globe like segments of an orange peel radiating from the north celestial pole, were named for prominent asterisms that lay within their boundaries. Specifying the lunar mansion of an object in the sky was akin to specifying the time zone of a location on Earth. In addition, each mansion carried its own particular astrological coloration, which could be invoked in reporting on events in the heavens. Locating a celestial event by its lunar mansion was not as precise as noting how far it was from a particular asterism, for the lunar mansions were larger. But the court astronomers would sometimes find it more astrologically meaningful to refer to a mansion rather than a nearby asterism or a particular star. An ideal of scientific accuracy was of course undreamed of at the time.

Nevertheless, in many cases the ancient reports are quite specific and remarkably precise. The past positions of the planets, which we can calculate today with great precision, provide one convenient way of checking which chronicles can be trusted. Certain Chinese records, for instance, mention the planet Venus entering the "mouth of Nan-tou," an asterism of six stars in the constellation of Sagittarius arranged somewhat like the Big Dipper. Using computers, we find that "Nan-tou" is just where Venus was located. On one date, October 24, 125, for example, Venus was moving swiftly through the gap between two stars that marked the "mouth" of the constellation. Two days earlier or later would clearly not have fit the description of the Chinese text.

Scholars agree that many Chinese records provide trustworthy reports of celestial phenomena. But were "new stars"—supernovae in particular—noted by the imperial astronomers? Indeed they were. With characteristic perceptiveness, the Chinese recognized three different classes of new stars that occasionally appeared among the old familiar ones. "Hui-hsing," roughly translated as "broom stars" or "sweeping stars," were most likely comets, characterized by a noticeable tail and usually a perceptible motion through the constellations. "Po-hsing," "bushy" or "rayed stars," probably referred to comets without tails. Such tailless comets include ordinary comets oriented so that the tail is not seen, as well as comets that never produce much of a tail, either because they never get close enough to the sun to be heated sufficiently, or because their surface material is not easily evaporated. Finally, "k'o-hsing," which means "guest star" or "visiting star," was used to designate unheralded dots of light in the sky. The "k'o-hsing" mentioned in the ancient chronicles include both common novae and supernovae, along with occasional misidentifications of comets or meteors. If used with caution, the ancient reports of "k'o-hsing" can be put to thoroughly modern uses.

HOW TO RECOGNIZE A SUPERNOVA

How are we to separate sheep from goats, to decide which "k'o-hsing" are common novae, which are supernovae, which are comets, and which are pure fancy? How can we be sure that the celestial event described in a dynastic history was not "cooked" to favor a particular interpretation or to flatter an emperor?

Some observations can be rejected out of hand. If the "guest star" is described as moving among the stars, it must be within our solar system, a comet, likely, or perhaps a bright meteor. Shape is also a clue, for stars appear pointlike. If the object is "hairy" or "broom shaped," or shows a tail, then no matter what the chronicles call it, it was a comet. If the object is described as "big," however, this only means it was bright. Before telescopic observations revealed that stars were unmagnifiable, it was generally accepted that the brighter stars appeared bigger.

Other reports in the ancient records are simply implausible. They may mention a configuration of planets that could not have occurred at a particular time, or they may locate a particular event in a constellation that was only above the horizon during the daytime. Occasionally the records have simply been dated incorrectly; in many cases the sightings we have in the written records were taken from other sources, years after the event, and inadvertently garbled in the transmission.

When all these obvious false alarms have been recognized, we are left with a small number of likely "new" stars among the ancient chronicles. David Clark and F. Richard Stephenson, for instance, listed 75 candidates in an influential 1977 study of the historical records. Of these 75, only a few are likely supernova candidates. The rest include the common nova outbursts, as well as a few sightings of slow variable stars caught near the peak of their regular cycles. Which are the real supernovae?

One clear and convincing clue is the length of the outburst. Common novae brighten rapidly and fade within a few days or weeks, whereas supernovae linger for months. A recently recorded common nova, for instance, known as Nova Cygni 1975, reached 2nd magnitude (about as bright as the stars in the Big Dipper) on August 31, 1975. Almost directly overhead in the early evening, it was a sight that could not be missed, even by casual stargazers. I recall rushing out to see it as soon as I heard the news from a friend, and spotting it immediately. It was good that I didn't waste time. Within a week it had faded by almost a factor of 40 and hovered near the naked-eye limit, about magnitude 6. Astronomers in cloudy climates scarcely had a chance to see it in its glory.

Supernovae, however, hang on and on. The supernova of 1006, as we noted, was visible for over 2 years. As I write this chapter, exactly half a year after the first sighting of the 1987 supernova, it is only slightly fainter than it was back then, still clearly visible to southern hemisphere observers. Thus any ancient report of a bright, stationary star that remains visible for several months (Clark and Stephenson use 50 days as a benchmark) is a likely supernova, whereas one visible for only a few weeks is probably a common nova.

Finally, most supernovae in our galaxy should be seen close to the luminous band of the Milky Way. Common novae can occur nearly anywhere in the heavens (this is because, being more feeble than supernovae, they must be close to us to be seen, and the relatively nearby stars surround the sun uniformly in all directions). But supernovae that are detected in our own and in similar spiral galaxies seem to be located mostly in the flat disks of the galaxy, and thus should be seen where the concentration of stars is the greatest, near the Milky Way.

Selecting from the Chinese records of long-lived new stars only those candidates that lie close to the Milky Way, we are left with about a dozen. At this point we apply one final test: can we still see the remaining debris from the stellar explosion in the location of the reported guest star? For thousands of years after the light from the supernova explosion has faded, telltale bits of gas may still be seen speeding away from the site. In some cases the supernova remnant, heated by its interaction with surrounding interstellar matter, continues to emit faint light that can be detected from Earth. It usually appears as a shell or arc of wispy material roughly centered on the position of the original blast. Finding such a shell is a clear indication that the ancient guest star was a supernova.

Some supernova remnants can be photographed using optical telescopes, but most cannot. The disk of the Milky Way is filled with intervening dust and gas, a sort of interstellar smog, that dims the faint glow of more distant objects. In addition, the light generated by the supernova remnant, for various reasons, is often too weak and diffuse for easy detection. Yet though its visible light may be undetectable, other signals from the supernova remnant may still reach us. Frequently, an expanding supernova remnant will emit a large fraction of its energy in the form of radio waves. This radio emission penetrates the interstellar dust as easily as TV signals penetrate the walls of your house. Thus, if our eyes were sensitive to radio waves rather than light, we could see the shells of otherwise invisible supernova remnants scattered around the Milky Way.

Modern astronomers have such radio-sensitive eyes. They are

radio telescopes, highly sensitive receivers closely related to the home satellite dishes that have been appearing in backyards recently. The most advanced radio telescopes produce "radio images" of celestial objects that are as crisply detailed as the photographs taken with large optical telescopes. To produce an image, the radio telescope is pointed at a particular direction in space and the incoming signal converted to a dot on a display screen; the stronger the signal, the brighter the dot. If the telescope is then pointed to an adjacent direction in space, and then another, and so on, a complete picture of a portion of the sky can be made. Using such methods, radio astronomers have detected well over one hundred sources that look like supernova remnants; in only a few dozen cases have corresponding remnants been photographed with optical telescopes. Even in these cases, the radio images usually show more intricate and extensive detail than the conventional photographs.

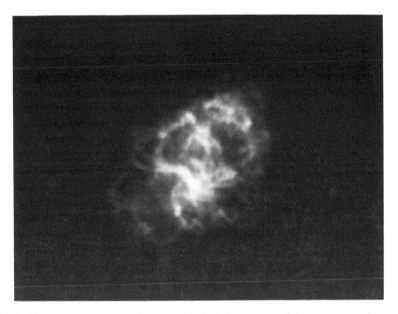

FIGURE 7: A radio image of the Crab Nebula, M1, remnant of the supernova of 1054. The image was produced from observations made between 1980 and 1982 by A. S. Wilson, D. E. Hogg, and H. H. Samarasinha using the Very Large Array (VLA) radio telescope in Socorro, New Mexico. (NRAO/AUI.)

In addition to radio emission, the hot gas in supernova remnants also emits X rays. Because Earth's atmosphere blocks X radiation from reaching the surface, astronomers were not able to make much use of this avenue of study until the late 1970s, when an X-ray telescope called the Einstein Observatory was placed in orbit around Earth. During its lifetime it produced a collection of detailed X-ray images of the sky that provide a rich source of information for astronomers. Among the X-ray images from the Einstein telescope are numerous ones of supernova remnants; these, too, can be used to confirm the reports of ancient supernovae.

When we examine all the records and consider all the evidence, just eight of the Oriental guest stars seem to correspond to identifiable supernovae. These were the guest stars first sighted in the years 185, 386, 393, 1006, 1054, 1181, 1572, and 1604. Each has a distinctive character when examined in the light of what we know today.

FIRST MILLENNIUM SUPERNOVAE

The Chinese chronicles contain just a single reference to the supernova of 185. It appears in the dynastic history of the Later Han dynasty, written around the beginning of the 5th century. There we read that on December 7, 185, "a guest star appeared in the midst of the constellation Nan-men; it was as big as the half of a bamboo mat and showed the five colours in turn, now beaming, now lowering. It diminished in brightness little by little and finally disappeared about July of the following year." The text goes on to elaborate the dire consequences of the new star: "When we come to the sixth year, the governor of the metropolitan region, Yuan-Shou, punished and eliminated the middle officials. Wu-kuang attacked and killed Ho-miao, the general of chariots and cavalry, and several thousand people were killed."

All indications are that this was a true supernova. When first sighted, it would have appeared from China not far above the horizon in the brightening sky of dawn. For it to have been seen at all, then, it must have been an exceptionally brilliant object, almost as bright as the quarter moon (magnitude -8 most likely). That it remained visible for almost two years would seem to rule out a nova, unless it was exceptionally close to Earth. Supernovae produce more light than normal novae, but this new star appeared so bright that even if it was a supernova it must have been situated in our general neighborhood of the Milky Way, within a few thousand light-years at the most.

From the brief description we have, we can still determine rather precisely where to look for the remnant of the supernova. Nan-men, the "Southern Gate," is an asterism that occupies a relatively small region of the southern sky, and includes two bright stars known today as Alpha and Beta Centauri. Between the two stars, a bit closer to the brighter of the pair, is a ragged ring of gas roughly equal in apparent diameter to the disk of the full moon. It shows up best in radio images and X-ray photographs. Optical telescopes detect just a trace of light from the gas, a few concentric arcs and filaments that look like the last ribbons of smoke from a cigarette. This fragmented shell, known as RCW 86, is almost certainly the remnant of the blast of 185, a clear indication that the Later Han astronomers were reliable observers of the sky.

The case is not so clear-cut for two guest stars that appeared in rapid succession in the 4th century, during the last part of the Chin dynasty (265–420). The first, in 386, was located in the asterism of Nan-tou, the Dipper-shaped grouping in the modern constellation of Sagittarius. Considerably fainter than the 185 guest star, it remained visible, according to the historical record, for only three months. It is possible that the star of 386 was faint because it was merely a common nova, but it also is possible that it was a true supernova much farther away than the one of 185. If so, then there should be a distant supernova remnant in the general vicinity of the asterism in Sagittarius. Though Nan-tou covers a large region of the sky, making a precise identification difficult, there is one remnant, G11.2–0.3 in the radio source catalogs, that meets this description. The most recent radio images, obtained in 1984 and 1985, show it as a clumpy ring of gas, roughly the same size and shape as RCW 86. Still, the vagueness of the ancient record casts some doubt on the identification.

Just seven years later, in 393, a guest star appeared for nearly eight months within the asterism Wei—the "Tail of the Dragon," a prominent group of stars in the modern-day constellation of Scorpio. We are certain that the 393 star was a supernova—no other object would have been visible for such a long time—but there are several possible supernova remnants in the vicinity of the reported position. Without additional details we cannot decide which remnant is the one whose birth was recorded in 393.

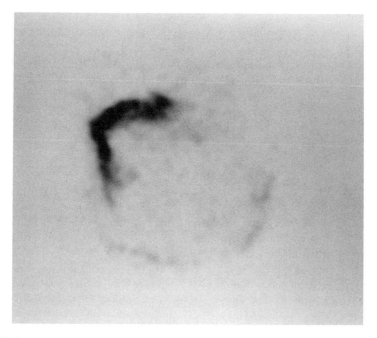

FIGURE 8: An X-ray image of RCW 86, remnant of the supernova of 185, taken by the Einstein X-ray satellite. The image is printed as a negative, with the dark parts representing places where the remnant is bright. This is common practice in astronomy, making it easier to see faint details. (Courtesy of Fred Seward, Harvard–Smithsonian Center for Astrophysics.)

From the ancient records, then, it seems that two, and probably three supernovae were seen in the first four centuries A.D. Two of the reports, unfortunately, provide us with little information. But the fourth recorded supernova, which occurred after a hiatus of 600 years, was to leave an extensive record in the historical literature. Even Europeans, who normally ignored sudden changes in the heavens, took note of this herald of the second millennium.

THE SUPERNOVA OF 1006

The new star that appeared in the far southern sky on May 1, 1006 is the brightest supernova on record. At maximum brilliance it probably reached magnitude −9.5, brighter than the quarter moon, bright enough, in fact, to cast shadows on the ground at night and to be visible against the blue of the sky by day. So outstanding was its brilliance that it must have drawn the attention of the most casual watchers of the heavens. And indeed, where earlier supernovae are only mentioned in one or two places in the historical literature, this one was recorded by virtually every literate civilization in Europe, the Middle East, and the Orient. David Clark and F. Richard Stephenson, in their exhaustive survey, *The Historical Supernovae*, list more than 20 accounts that range from official histories to diaries of contemporary observers.

Looking at the supernova of 1006 must have been like standing in the beam of a distant but powerful searchlight. Surely the observers of the time would have been astounded by a light that was simultaneously so intense and so starlike, for they would never before have seen anything like it, either natural or manmade. (The moon was comparably bright, but its light was spread out over its disk, not concentrated in a point.) It is not surprising that the records frequently blend a sober description of the star with wonderment and speculation as to its source and meaning.

In Cairo the scholar Ali ibn Ridwan, observing the new star, was so impressed that he recounted his observations in a commentary to a book of Ptolemy he edited years later:

> I will now describe a spectacle which I saw at the beginning of my studies. This spectacle appeared in the zodiacal sign Scorpio, in opposition to the sun. The sun on that day was 15 degrees in Taurus and the spectacle in the 15th degree of Scorpio. This spectacle was a large circular body, 2 1/2 to 3 times as large as Venus. The sky was shining because of its light. The intensity of its light was a little more than a quarter that of moonlight. . . . Because the zodiacal sign Scorpio is a bad omen for the Islamic religion, they bitterly fought each other in great wars and many of their great countries were destroyed. Also many incidents happened to the king of the two holy cities [Mecca and Medina]. Drought, increases of prices, and famine occurred, and countless thousands died by the sword as well as from pestilence. At the time when the spectacle appeared calamity and destruction occurred which lasted for many years.

Ali ibn Ridwan's reference to the size of the new star was, of course, consistent with the old practice of assuming that brighter stars were larger stars. And he was clearly struck with the astrological implications of the "spectacle." Still his description of the position and brightness of the object are quite intelligible in modern terms.

European chroniclers were seldom as careful or as astronomically knowledgeable as this. Typical were the remarks of Alpertus of Mertz in France, who was alive at the time of the supernova. For some reason, he erroneously gave the date of the star as 1005 when he wrote:

> A comet was seen in the southern part of the sky with a horrible appearance, emitting flames this way and that. In the following year, a most terrible famine and mortality took place over the whole Earth, with the result that in many places on account of the multitude of the dead and the weariness of those who were burying them, the living, still dragging their breath and struggling with what strength they had, were overwhelmed along with the dead.

The position of the object is vague, its appearance is described in terms more rhetorical than scientific, and the star is mistaken for a comet. But the description, even its association with plague and famine, follows the narrative of other chronicles, such as the record of the St. Gallen Monastery quoted in the Preface of this book. We are certain it describes the new star of 1006.

Each of the ancient records, taken singly, specifies a rough position for the supernova of 1006. That's not much help to radio astronomers looking for a small remnant in a big sky. Clark and Stephenson, however, have examined all the records from China, Europe, and the Near East to determine their common region of overlap. The result is a position of remarkable precision.

The observation at St. Gallen, that the star was seen "in the inmost extremities of the south, beyond all the constellations which are seen in the sky," is particularly informative. From any observation point near the Monastery of St. Gallen, the Alpenstein mountains trace a ragged line across the far southern horizon. Only a limited part of the sky is visible just above the mountains, "in the inmost extremities of the south." It is a small strip of the celestial sphere, about five degrees in width, running roughly east and west.

The account of Ali ibn Ridwan that the "spectacle" was in the 15th degree of Scorpio limits the position to an even thinner strip running roughly north and south, which intersects the strip seen from St. Gallen. Add to these limits the Chinese observations, which place the "guest star" in the modern constellation of Lupus, and we are left with just a small patch of sky, not much larger than the full moon, enclosing the probable position of the supernova.

And there we find its remains. At the northeast edge of the

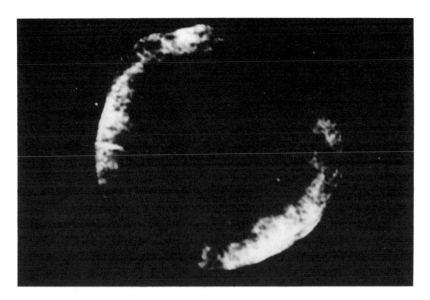

FIGURE 9: A radio image of the remnant of the supernova of 1006. Though it looks like a ring, the three-dimensional structure of such remnants more nearly resembles a shell with most of the emission coming from its outer edges. The image was produced from observations made in 1983 and 1984 by Stephen P. Reynolds and Diane M. Gilmore using the Very Large Array. (NRAO/AUI.)

region selected by Clark and Stephenson, the radio catalogs list a small circular radio source, known as PKS 1459–41. It is clearly the remnant of a supernova, and it appears to be about a thousand years old. In the 1970s astronomer Sidney van den Bergh, using a large telescope in the mountains of Chile, photographed a few faint optical wisps from the 1006 remnant. Several years later, the Einstein satellite was able to return an X-ray image of it. There's not much left of the bright star that foretold bloodshed and famine. A thin shell of gas is all that we can see of the brilliant supernova that ushered in the second millennium.

THE SUPERNOVA OF 1054

In the Municipal Museum in Suzhou, China, there is a star map, inscribed in stone, that stood for many years outside the Confucian

Temple of the city. Almost 1500 stars are inscribed on the map, which dates to 1247, and in most cases their positions are accurate to within a few degrees when compared to modern maps. Yet there is one mark on the map, a small circular indentation just to the northwest of the star T'ien-kuan, in the modern constellation of Taurus, that does not correspond to any star seen on current maps of the sky. It is a rendering of the guest star of 1054, not the brightest, but surely one of the most intensively studied, of all the early supernovae.

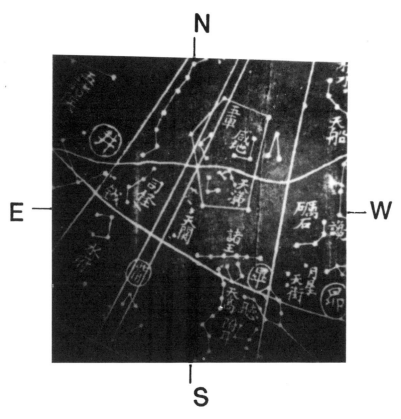

FIGURE 10: A section of the Suzhou star map. The mark identified as the guest star of 1054 is located at the center of the photograph, just below and to the left of the large polygon of stars. The lowest star in that polygon is known today as Beta Tauri. This is a photograph of a rubbing taken from the stone chart. The rubbing is at the National Academy of Sciences in Washington, D.C. (Courtesy of Kenneth Brecher.)

The annals of the Sung dynasty report that the star first appeared early in the morning of July 4, 1054, near the asterism T'ien-kuan (the written account says southeast, rather than northwest, but there is general agreement that this was a mistake in the recording of the position rather than a discrepancy in the observation), and that the star remained visible for more than a year. Other observers noted that the star was as bright as the planet Jupiter and that it could be seen distinctly during the daytime. This would make its magnitude at maximum about −4, about six times brighter than the star Sirius. The 1006 supernova was probably about 100 times brighter that this, but the 1054 event, which appeared much higher in the sky as seen from northern latitudes, would still have been impressive.

Though the supernova of 1006 was widely reported in Europe, there is not a single European record of the supernova of 1054, save for a passing reference by a Christian physician from Baghdad, Ibn Butlan, who lived in Cairo and Constantinople between 1052 and 1055. Unless the entire continent of Europe, from the shores of the Mediterranean to the northern limits of civilization, was covered with clouds for three or four months, we can assume that many people in Europe saw it, too. Why did they not write about it?

One school of thought is that Europeans did not recognize new stars as being celestial phenomena. According to the teachings of Aristotle, the heavens beyond the moon were changeless, and therefore anything that changed in the sky was an atmospheric disturbance, of no more note than a bolt of lightning or a puff of cloud. (We discuss this at greater length in the next chapter.) In this regard historian of science George Sarton has accused medieval Europeans of "prejudice and spiritual inertia connected with the groundless belief in celestial perfection." Sarton's accusation is not altogether convincing, however: that same groundless belief in perfection did not prevent Europeans from keeping records of the supernova of 1006, less than 50 years earlier.

Another school of thought blames strife within the Church for the absence of astronomical records. There were clouds over Europe when the supernova appeared, but they were storm clouds of a spiritual kind. The Eastern and Western divisions of the Catholic Church, always at odds over matters of doctrine and authority, had reached the point of total disaffection. In mid-July, two weeks after the star exploded, the Roman Pope Leo IX formally excommunicated Patriarch Michael Cerularius and the entire Eastern Church. (The ostensible reason was a disagreement about the wording of the Apostle's Creed.) The break between Rome and Constantinople is commonly known as the Great Schism. At such a time, the wisest course was probably to keep mum about any omens in the sky. Astrology itself was seen as a danger to the spiritual hegemony of the Church, and the fact that such a powerful portent had appeared immediately before such a profound dispute might have prevented anyone of intelligence and discretion from writing about the new star.

There are indications that Oriental astronomers were not the only ones who saw the star. In 1054 the Anasazi, ancestors of the present-day Pueblo people, had established a complex and flourishing civilization in the American southwest. At Chaco Canyon, a depression that runs roughly east and west through the weathered sandstone cliffs of northwestern New Mexico, they constructed expansive, intricate structures of masonry, linking their settlements to each other and to outlying regions with hundreds of miles of broad, straight roadways. Pueblo Bonito, the largest of the Chaco ruins, is a semicircular building covering three acres of ground, which, when new, rose to four stories and contained eight hundred rooms. According to archaeologist Neil Judd, who excavated it in the 1920s, "No other apartment house of comparable size was known in America or in the Old World until the Spanish Flats were erected in 1882 at 59th and Seventh Avenue, New York City."

The Chaco people seem to have been unusually attentive to the sky. They built numerous ceremonial sites throughout the Canyon that clearly were used to record the position of the sun on important days of the year. At Fajada Butte, a prominent sandstone outcropping near the southeast end of the Canyon, the Anasazi devised an ingenious method of calendar keeping. Three thin slabs of rock lean against the Butte not far from its top, admitting a shaft of sunlight that strikes the cliff face behind them. There the Chaco carved a spiral into the sandstone. At midday, and only at midday, on the day of the summer solstice (the first day of summer), a "dagger" of sunlight passing between the leaning slabs pierces the center of the spiral. The "sun dagger" reveals a keen imagination and a close attention to the motion of the sun throughout the year. Most likely the Anasazi watched the moon, planets, and stars as well.

At the other end of the canyon, painted on an overhanging slab of sandstone, is a depiction of a crescent moon with a starlike object—a circle with radiating rays—not far from one horn of the crescent.

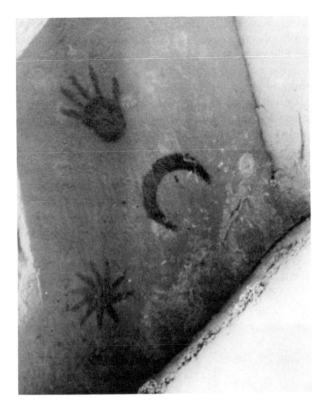

FIGURE 11: Pictograph on a sandstone surface at Chaco Canyon, New Mexico depicting the crescent moon and a star. It is a probable representation of the supernova of 1054. (Courtesy of Kenneth Brecher.)

This, it has been suggested, represents the 1054 supernova that, on the morning of July 5, appeared just before dawn right beside the rising crescent of the moon. It is not hard to imagine a Chaco shaman, intent on the changing positions of the rising sun and moon, recording the sudden and startling appearance of a new object in the heavens by a painting like this.

Although astronomical symbolism is relatively rare in Native American pictorial remains (we have, unfortunately, no written records of these cultures), the combination of crescent moon and nearby star is widespread in the Southwest, which suggests that the 1054 supernova attracted the attention of many people there. The first of such markings was discovered in the mid-1950s by William Miller, then chief photographer at Palomar Observatory in California, and astronomer Helmut Abt, at White Mesa in northern Arizona. Since that time almost a dozen renditions of the possible 1054 sighting have been discovered from California to Texas. (But what of the much brighter 1006 event? It is possible that, because it never appeared close to the moon, it was not accorded the same attention. Or that, pictured simply as a star by itself, we cannot recognize it.)

The 1054 supernova is significant, not merely because we have such widespread records of its appearance, but because it marked the birth of the Crab Nebula, one of the nearest, brightest, and most carefully studied of the supernova remnants. The records seem clear on this, for the Crab Nebula (number 1 on Messier's list of nebulous objects) is the only remnant just to the northwest of Thien-kuan, the star we now call Zeta Tauri.

Unlike the 1006 supernova, optical photographs readily show the remains of the 1054 blast; it can even be seen with a small telescope,

FIGURE 12: The Crab Nebula, remnant of the supernova of 1054, photographed in visible light. (Lick Observatory photograph.)

just to the east of the horns of the constellation Taurus the Bull. Its appearance, whether in optical, X-ray, or radio telescopes, differs radically from the other remnants we've mentioned. There is no distinct circular shell. Instead, the main body of the Crab appears as a roughly S-shaped blob. Superimposed on the blob is a network of tangled filaments that presumably fill the main body of the nebula. A comparison of photographs taken at different times shows that the tangled filaments are expanding.

Because the Crab is such a distinctive object, it figures prominently in our understanding of the remnants of supernovae, to be discussed at more length in Chapters 7 and 8. There we shall see that the 1054 explosion marked not only the birth of a nebula but also the death of a star, whose remains, a dense, rapidly spinning neutron star, can also be detected at the position of the ancient records, near the center of the expanding tangle of gas in the Crab. That star, called the Crab pulsar because it emits regular pulses of radio waves, has also been enormously influential in the modern study of supernovae.

Thus, the 1054 supernova created two remarkable objects, a diffuse gaseous remnant and a dense stellar cinder, which were undreamed of by those who first saw it, in China, the Middle East, and North America. Nine centuries were to pass before the meaning of the midsummer guest star was to become clear.

THE SUPERNOVA OF 1181

Perhaps because it appeared in the constellation of Cassiopeia, high in the early evening sky, the guest star of 1181 was widely observed throughout the Orient. No doubt then, as now, people were fond of strolling after dusk, and anything easily visible would have received a wider audience than an object that only could be seen late at night, or close to the haze of the horizon. We know of reports not only in the Chinese dynastic histories, but also in official histories and even several diaries from Japan. Fujiwara Kenezane, a courtier of the Japanese imperial court, in an entry for August 7, 1181, noted the presence of the guest star near the asterism Wang-liang, and concluded that it was "a sign of abnormality indicating that at any moment we can expect control of the administration to be lost."

According to the contemporary Chinese histories, the 1181 guest star was observed for almost half a year, which indicates that it proba-

bly was a supernova; but it was surely not as brilliant as its predecessors of 1006 and 1054. The star of 1181 probably never rose
above magnitude 0, about a hundred times fainter than the 1054 star
at its brightest, and several thousand times fainter than the star of
1006. It still would have ranked among the two or three most brilliant
stars in the heavens, which helps explain the number of accounts
given by nonastronomers of the time.

It may be that the 1181 supernova was a bit more distant than its
predecessors, or located behind regions of obscuring dust that made
it appear fainter. But faintness is not its only peculiarity: the association of the reported positions with a cataloged supernova remnant is
less secure as well. There is an oval-shaped radio source, known as
3C 58 (number 58 in the 3rd Cambridge catalog of radio sources), that
appears close to the asterisms where the supernova was reportedly
seen. The glowing cloud of gas seems rather distant, about 25,000
light-years from us, which is consistent with the relative faintness of
the exploding star at maximum light.

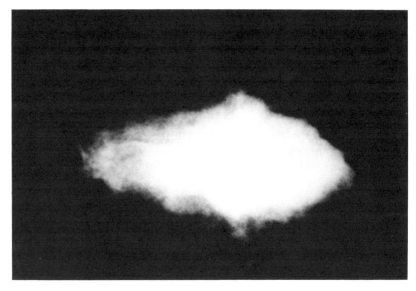

FIGURE 13: A radio image of 3C 58, the remnant of the supernova of 1181. The image
was produced from observations made in 1984 by Stephen P. Reynolds and Hugh D.
Aller using the Very Large Array. (NRAO/AUI.)

Even if we accept the identification with the supernova of 1181, 3C 58 is a bit of an oddity. Radio images fail to show a bright ring around the outside as is the case for the remnants of the supernovae of 185 and 1006. 3C 58 does resemble the Crab Nebula and a small number of other remnants that appear as filled-in blobs rather than hollow shells. Though it is younger than the Crab Nebula and the remnant of 1006, there is no optical remnant visible, even through the largest telescopes. Despite the large number of people who reported seeing it, the supernova of 1181 remains one of the more puzzling of the historical supernovae.

A THOUSAND YEARS OF WATCHING

An ancient tradition of diligent sky-watching in China made possible the discovery of half a dozen new stars in just under a thousand years. From these figures alone one can conclude that supernovae are far rarer sights than most phenomena in astronomy. On a clear night one has only to gaze skyward for a few minutes to catch sight of a meteor flaring across the sky. Bright comets appear a half-dozen times each century. Many of us have watched an eclipse of the sun or of the moon; and a fair number of us have seen several. But even if we were to scan the skies with the dedication of an Imperial Astronomer, fearful of incurring the emperor's wrath, we could scarcely hope to see a naked-eye supernova during the span of several lifetimes.

More than four centuries, in fact, were to pass before the Oriental records again noted the appearance of a brilliant k'o-hsing, a guest star, in the evening skies. This time Europeans were watching, too.

CHAPTER 4

Supernovae and the Revolution in Science

> And in these Constellations then arise
> New starres, and old doe vanish from our
>   eyes.
> —John Donne, *First Anniversary*

ARISTOTLE AND THE MEDIEVAL COSMOS

A new star in ancient times was a troubling sight. It violated the long-established order, raising doubts about the reliability and regularity of the heavens. Little wonder that Oriental astronomers and Arab scholars reported such appearances as signs of pestilence and war. Emperors sickened and died, the tide of battle turned, dynasties toppled, all supposedly in response to the changes in the sky. Such superstitions have lost their force with time, but looking back we still can marvel at the influence of two bright new stars, which medieval Europeans called novae, but which we recognize today as supernovae.

These two new stars, the first in 1572 and the second in 1604, brought intellectual change and spiritual unease, not plague and famine. They arrived at a time when Western thought was in turmoil. A new skepticism was challenging cherished philosophical and religious beliefs, and the structured, earth-centered cosmos of medieval times was gradually giving way to the impersonal, mechanistic world of modern science. The two novae, which hastened the establishment of a new science, thus signaled a break with a tradition of thinking

that went back almost 2000 years, to the time of the Greek philosopher Aristotle.

Up to this time, medieval thinkers, with few exceptions, zeal-
ously admired the imagination and scope of Aristotle's work. He
seemed to have studied and commented on everything, and scholars
devoted entire careers to translating, collating, and interpreting his
writings. Those on astronomy were particularly influential. In Aristotle's cosmos, the universe consisted of a series of nested concentric
spheres, not unlike the nested "Matryoshka" dolls produced by Russian folk artisans, where one wooden doll opens up to reveal a smaller doll inside, and yet another smaller doll inside that. The motion of
the nested spheres, according to Aristotle, determined the course of
all events of importance in the heavens.

Earth—a solid, dense, immobile globe—occupied the precise
center of the universe. Around it spun the outer spheres, ethereal
and crystalline, carrying with them the planets, sun, and moon. The
first sphere carried the moon, spinning about Earth once a month.
Beyond this were separate, larger spheres for Mercury, Venus, the
sun, Mars, Jupiter, and Saturn. An outermost eighth sphere, to
which the stars were attached, rotated once each day, carrying all the
inner spheres with it. The daily rotation of this outer sphere neatly
accounted for the rising and setting of stars and planets. Aristotle
required additional invisible spheres, a total of 55 in all, to complete
his system. The extra spheres functioned like the gearbox of a car,
coupling the rotations of the outer spheres to the inner ones and re-
producing, in rough form, the looping motions of individual planets.

This was no mere astronomical fancy, of consequence only to the
mathematically inclined. It was a complete cosmology that embraced
all of creation, both physically and spiritually. In Aristotle's system,
the behavior of things on Earth, even an event as mundane as the
falling of a stone or the flight of a sparrow, was tightly linked to the
overall structure of the universe. Above the sphere of the moon tran-
quility reigned, and the heavens were essentially perfect and changeless; they were close to divine. But the region below the sphere of the
moon was regarded as a region of turmoil and change; it was corrupt,
like the fallen beings who occupied it. Medieval thinkers thus regarded the heavens and Earth as separate but complementary parts of
a universal organism, whose functioning produced all the events we
saw around us.

Even though Aristotle's ideas are long out of fashion, it's not hard to understand the power of his arguments in explaining the world to 16th-century Europeans. All around us we see change. Creatures are born, die, and decay; mountains are eroded by wind and rain. "You can never step twice into the same river," wrote the philosopher Heracleides, despairing of ever achieving certain knowledge in a world that seemed to be in a state of unremitting flux. But to Aristotle there was an underlying cause of all this change, something that remained constant through all the flux and strife. Terrestrial substances, he claimed, were formed of mixtures of four fundamental elements: earth, air, fire, and water. Jumbled up in various proportions, they made up all the things that could be found below the sphere of the moon. Change was an effect produced when these elements were separated or rearranged. Each element, when released from a mixture, would seek a natural resting place in the universe. Water and earth sank down, air and fire went up. When a log was burned, for instance, it released air (smoke) and fire, which rose upward, leaving behind water (sap) and earth (ash).

I like to imagine Aristotle's account of this terrestrial realm as a mixture of the four elements, shaken together like oil and vinegar in a salad dressing. Left to themselves, the elements might separate out, forming concentric spheres of (from the center out) earth, water, air, and fire, just as layers of oil and vinegar separate in a bottle left to stand. But the terrestrial world remains far from this state of equilibrium; it's kept constantly mixed by the interactions of its various parts and by the turmoil raised by man and beast. Thus, we see imperfection and change all around us.

Above the sphere of the moon, however, the behavior of the universe was quite different. Heavenly bodies, Aristotle taught, were composed solely of a fifth substance, the "aether" or "quintessence," which was pure, changeless, and perfect. Left to themselves, heavenly bodies would circle around the center of the universe (Earth) forever, rather than rising or sinking like bodies made of the other four elements. A body moving in a circle would trace and retrace the same path over and over again—though it moved, nothing really changed. Furthermore, in substance as well as in motion, the pure, heavenly aether was eternal and incorruptible. Nothing was born, nothing died, and nothing altered above the sphere of the moon.

How then to explain the sudden appearance of a comet, trailing a long and fiery tail through the sky? The glitter of a meteor flaring briefly into existence? The visit of a "guest star"? The silent glow of the aurora? Mere atmospheric phenomena, huffed the Aristotelians. Unlike the motions of the heavens, which reflected an eternal, repetitive order, these unexpected changes in the sky were events that occurred below the sphere of the moon, and were caused by the changeable nature of earthly matter.

We saw changes in the nighttime sky, said the Aristotelians, because fiery substances (i.e., things made of the element fire) occasionally condensed in the upper atmosphere, close to Earth on a cosmic scale. They had no more significance to the celestial order than a bolt of lightning or a rainstorm. This belief, as we noted in an earlier chapter, may be why Western observers of supernovae (and there must have been many) long missed the astronomical significance of what they had seen. The word *meteor* itself, though now considered astronomical, once applied to any change seen in the sky. It owes its origin to the Aristotelian view that lights in the sky were only atmospheric in nature. Much of what we call astronomy today, in other words, was mere "meteorology" to the Aristotelians.

Medieval scholars turned to Aristotle for the final word on anything that was in doubt. They quoted him for support like a fundamentalist preacher quotes the scriptures, and invariably found a way to interpret his writings to support any point of view. Yet medieval thinkers found Aristotle's cosmos satisfying not only because of its power to explain (or explain away) so much of the natural world, but also because it meshed so smoothly with the Christian view of God and humanity. The terrestrial realm, full of change, was seen as a morally corrupt region, the site of the Fall from Paradise. Rising through the spheres of the heavens, one drew closer and closer to the perfect presence of God, who ruled just outside the final sphere of the stars, giving everything a whirl now and then to keep the universe going. Man was at the bottom of the divine hierarchy, but also at the center of creation. Through divine grace we could rise to the heavens. The Aristotelian universe, to the medieval mind, was a comforting, if not comfortable, place to live. All creation was infused with meaning.

It was not to last. The foundations of the Aristotelian system, shaky to begin with, were showing definite signs of age by the 16th and 17th centuries. Scholars challenged one, then another of Aristotle's assertions. At the same time, religious reformers like Luther questioned the spiritual power of the Church. They encouraged a spirit of inquisitiveness, a growing feeling that individuals could discover truth by logically considering facts, rather than by appealing to established authority. More conservative souls saw blasphemy in all this questioning, and feared a world cut free from the old beliefs. It was in this unsettled climate that Nicolaus Copernicus, a canon of the Frauenburg Cathedral on the Baltic coast of Poland, set forth a new view of the universe that ultimately loosened the bonds of Aristotelianism and ushered in a revolution in science.

Copernicus was well known and well respected for his astronomical knowledge. But fearful perhaps of ridicule or censure, he did not publish a full account of his ideas until the very last opportunity. In 1543, as he lay on his deathbed, Copernicus oversaw the publication of his lifework, *On the Revolutions of the Heavenly Spheres*. The sun, he proposed, was the rightful center of the universe, with the planets, including Earth, circling it. To be sure, the planets were attached to spheres just as in Aristotle's universe, and an outermost sphere carried the stars. But the very notion that Earth was one of many planets challenged the old division between a corrupt sublunar region and a changeless heaven, the comforting image of humanity at the center and God all around. If Earth could move through space, could there perhaps be other changes in the heavens? Could there be other earths? Other intelligent beings? Other saviors? The possibilities were profoundly disturbing. Copernicus's book became controversial, discussed by many, embraced by only the boldest and most independent thinkers. But few could muster clear, persuasive evidence to favor one view of the universe over another. A brilliant nova changed that state of affairs.

Late in 1572 a new star unexpectedly appeared in the constellation of Cassiopeia, the first nova seen in the West in nearly four centuries. Many saw it, speculated on its meaning, and wondered how it fit into the accepted scheme of things. Was it just another ball of atmospheric fire? Or a sign from God? One astronomer went beyond idle speculation, measuring the position of the new star systematically and with unprecedented precision. Tycho Brahe, then a young man unsure of his calling, was to become the last great astronomer of antiquity, and, thanks in part to the star, the first great astronomer of the modern age.

TYCHO AND THE NOVA STELLA OF 1572

Tycho Brahe was the younger of twin sons born in 1546 to a family of Danish nobility. His elder brother died shortly after birth, and soon thereafter Tycho was abducted by his uncle Jörgen who, childless and lonely, spirited the infant to his nearby estate. His parents were understandably anguished and upset. Yet for some reason, unclear in the few accounts we have, they reconciled with Jörgen after Tycho's mother gave birth to another son the following year, and Tycho became a legitimate member of his uncle's family from that time forward.

Tycho's later life was no less remarkable and eventful than his birth. His artificial nose has become as famous as George Washington's false teeth. It was cast of gold and silver, enameled to resem-

FIGURE 14: Tycho Brahe. (Yerkes Observatory.)

ble flesh, and attached by some type of ointment or paste. Tycho had it made after his real nose was cut off in a college duel, reputedly over a disagreement on some point of mathematics. His youthful flamboyance was tempered only slightly with age, and the astronomer remained both a pride and an embarrassment to the Danish court. He could be vengeful and blustering, treating his serfs with disdain and his professional rivals with rancor. In 1601, during a heavy bout of drinking at the house of a friend, he declined to leave the table, out of a sense of etiquette. He was taken home suffering great pain, and eleven days later he died of a bladder infection.

Yet as an astronomer Tycho was a towering intellect and a tireless worker. Schooled in Denmark and Germany, he abandoned the pursuits of medicine and law (the usual professions for gentlemen of the day) to study astronomy and alchemy. He became so well known for his special knowledge that, in 1576, King Frederick of Denmark granted him an extravagant stipend, along with a small island, Hveen, just off the shore between Copenhagen and Elsinore. (Hveen today is a part of Sweden.) There Tycho established a palatial home, Uraniborg, along with an observatory equipped with highly accurate sighting devices of his own design.

The island of Hveen became the center of astronomical activity in 16th century Europe. Tycho, cocksure and imperious, supervised the observations and publications of the establishment. He entertained scholars and noblemen, demonstrating his unique inventions to all who were interested, and regularly presiding at communal dinners, during which his dwarf Jeppe sat at his feet to receive table scraps. A regular staff of observers kept his instruments in constant use, and his books on astronomical equipment and techniques, though soon to be superseded by the invention of the telescope, established a standard of excellence for astronomers of the time.

Astronomers in the 16th century could only measure the angles between objects in the sky, using sighting devices like the sextant or the quadrant. These were angle-measuring devices, like giant protractors, with movable arms that could be aligned with the stars just as the sight on the barrel of a rifle can be aligned with a target. Sextants could measure separations up to sixty degrees (a sixth of a circle), and quadrants, angles up to ninety degrees (a quarter of a circle). The naked eye was the light detector, pen and paper the recording device. Making observations with these instruments was difficult and exacting work, but Tycho did it better than anyone else.

For a quarter of a century he kept a continuous record of the positions of the planets among the stars, more complete and precise than any that had preceded it.

Tycho's underlying purpose was to disprove Aristotle, but he was not an ardent Copernican. He championed his own pet theory of the universe in which the sun went around Earth and all the other planets orbited the sun. The facts, however, proved him wrong. In the hands of his assistant, Johannes Kepler, Tycho's observations provided the framework for the more modern notion that the sun, not Earth, was the dominant body in the solar system. Kepler used Tycho's records to show that the planets, including Earth, moved around the sun, and that their orbits were ellipses, not the perfect circles of Aristotle. A half century afterwards, Isaac Newton explained Kepler's results mathematically, showing how the sun's gravitation shaped the orbits of the planets. Newtonian theory replaced the crystalline spheres of Aristotle with the mathematical laws of physics, and the solar system began to be seen as a smoothly running clockwork wheeling in infinite space, not a divine organism centered on Earth.

All this lay in the future when, on the evening of November 11, 1572, Tycho Brahe, still a young student, was returning home from a day in the alchemical laboratory he shared with his uncle Steen Bille. Glancing at the sky from force of habit, he was brought up short—something was glaringly out of place. There was a new star in the northern sky, more brilliant than the planet Venus, right next to the middle of the "W" traced out by the bright stars of the constellation Cassiopeia. To a seasoned sky-watcher, it was as startling a sight as seeing a mustache drawn on the Mona Lisa.

Unable at first to believe his eyes, Tycho asked the servants who accompanied him whether they saw what he was pointing at. They did. Still amazed, he stopped some passing peasants to make sure that he and his men were not suffering a mass delusion. Everyone saw the star, of course. High in the sky, brighter than any ordinary star, it was unmistakable, though we may wonder whether any of those present, other than Tycho, knew enough about the constellations to be astonished by what they saw.

Tycho was not the first to spot the interloper, however. Observers in Sicily and Germany saw it five nights earlier, on the 6th of November, and Michael Maestlin, a noted astronomer who many years later became Kepler's mentor, observed it on November 7. It

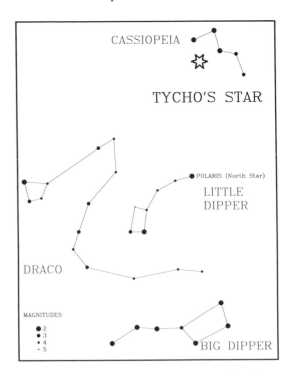

FIGURE 15: Chart of the northern sky showing the position of Tycho's star, the nova of 1572, now recognized as a supernova. (Courtesy of Ken Croswell.)

had brightened into visibility sometime after November 2, according to the testimony of Hieronimus Munosius, a professor at the University of Valencia. On that night, he recalled, he was teaching an outdoor class on star identification, and would surely have noticed the nova (as he did several nights later), had it been there.

Most observers, however, were content to marvel at the new star, making only cursory efforts to estimate its position and brightness. Maestlin, clever and resourceful, measured the position of the star by stretching a string held at arm's length between two pairs of stars in Cassiopeia and noting where the nova lay along the line joining them. Tycho was even more systematic. He had recently acquired a new sextant, a handsome instrument consisting of two hinged walnut arms along which objects in the sky could be sighted, the angle between the arms being measured on an attached scale graduated to the nearest minute of arc. Using the sextant he carefully measured the separation between the nova and the nine bright stars of Cassiopeia. He repeated the observation several times during the night, and continued to monitor the position of the nova for the many months that it remained visible.

Tycho's effort may seem excessive—one set of measurements, or two at most, would seem sufficient to determine a star's position. But Tycho was intent on testing one of Aristotle's cherished assertions— that new stars, appearing and disappearing as they did, were sub- lunar events. By measuring the precise position of the nova during the course of the evening, he hoped to determine whether the object moved against the background of more distant stars. This would give him a way of telling how far away the nova was. During the course of an evening, he knew, the moon exhibits a so-called "parallax" shift as the line of sight from an observer to the moon changes with the rotation of Earth (or the rotation of the heavens, to one trained, as Tycho was, in Aristotelian astronomy). The nightly parallax shift of the moon among the stars (sometimes called the "equatorial paral- lax") amounted to several times the moon's diameter, and could easi- ly be detected with Tycho's instruments. Any object closer to Earth than the moon would display an even larger nightly parallax. But the nova of 1572, Tycho noted, showed absolutely none. It never budged, as fixed as if it were nailed to the sphere of stars, until, 18 months after it had first been sighted, it faded to invisibility. The nova, Tycho concluded, was far more distant than the moon.

During this time, Tycho also kept a careful record of the bright- ness and color of the star. When first seen, he wrote, it was an intense white, bright as the planet Venus (about magnitude −4, in modern terms). Even on overcast nights it could still be seen as an indistinct glow through the clouds. By December it had faded a trifle; Tycho compared it to the planet Jupiter, and in January it was just a bit brighter than stars of the first magnitude. The next summer it was at about third magnitude, roughly equal to the other stars in the con- stellation of Cassiopeia, and its color had changed to a pastel red. By March of the following year it had disappeared entirely. Nearly four hundred years later, astronomer Walter Baade was able to use Tycho's observations (along with those of Maestlin and other reliable observers) to plot an accurate graph of the brightness of the 1572 nova. The light curve (as astronomers term such graphs) bears a striking resemblance to what modern astronomers now call a Type I supernova, a testimony to Tycho's skill as an observer. The precision of Tycho's position measurement, in addition, later enabled radio astronomers to locate the remnants of the exploded star with relative ease. Optical and X-ray astronomers have also been able to detect the supernova remnant; it clearly resembles the shell of gas left by the 1006 supernova.

Most of the 16th-century astronomers who saw the nova showed far less skill and far more credulity than Tycho. Some claimed to have measured a parallax for the star—as was to be expected if, in accord with the Aristotelian orthodoxy, it was a sublunar puff of fire. Their instruments were, by and large, crude compared to Tycho's, which probably explains the discrepancy. Others accepted the celestial na-

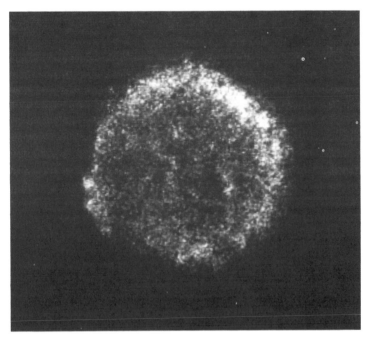

FIGURE 16: An X-ray image of the remnant of the supernova of 1572, taken by the Einstein X-ray satellite. (Courtesy of Fred Seward, Harvard–Smithsonian Center for Astrophysics.)

ture of the star, but suggested that it was an old, faint star made suddenly visible by some change in the intervening atmosphere of Earth. And, as usual, the astrologers and the cranks had a field day with the nova, capitalizing on the general curiosity about an object that had no business being where it was. Superstitious interpretations filled the popular press. It was an optical illusion; it was a reappearance of the Star of Bethlehem; it was a sign of the Second Coming of Christ. In one pamphlet German painter Georg Busch identified the nova as a comet. According to Busch, the comet was composed of a sort of noxious gas generated by human sin, which floated heavenward until ignited by the wrath of God. As it burned, Busch wrote, the comet became a prolific celestial polluter, showering its effluence widely over Earth and thereby causing "pestilence, Frenchmen, sudden death, bad weather, etc."

Such nonsense rankled the young Brahe, but he was at first hesitant to write about the nova. In his day noblemen were not supposed to dirty their hands with pen and ink. Yet he spoke freely with fellow astronomers and with the intelligentsia of Copenhagen, stressing the trustworthiness of his observations and his conviction that the star was neither an omen nor a meteoric emanation. At his friends' insistence he published, in 1573, a small volume, *De Nova Stella*, only 52 pages in length, which presented his observations of the star and his estimation of its astrological meaning and astronomical significance. Had he published nothing else, his place in history would have been assured. The nova of 1572 has ever since been identified with Tycho's name.

Tycho was apparently utterly convinced by his measurements of the star that something was seriously amiss in the Aristotelian system. Because his star showed no motion, it must lie beyond the sphere of the moon, probably as far away as the eighth sphere of the fixed stars. Yet it had changed. When first seen it was more brilliant than a 1st-magnitude star. (Tycho also thought that it was more than a hundred times the size of Earth. Like other astronomers of his time, he erroneously believed that brighter stars were larger stars.) It had faded and turned redder over time. This was bad behavior for something made of heavenly stuff.

The nova reinforced Tycho's growing dissatisfaction with orthodox astronomy. His faith in Aristotle had already been shaken almost a decade earlier when he had noted that the Alphonsine Tables, the standard almanac of planetary motions calculated by Aristotelian methods, gave dates of celestial events that were sometimes several days in error.

Nevertheless, he realized that more evidence of the shortcomings of Aristotelian astronomy was needed to bring others around to his way of thinking. One book on one nova was not enough. A continuous and reliable record of the positions and motions of the stars and planets was needed to resolve the issue. Thus, the nova of 1572 was a vocation, a calling from heaven that inspired his dedication to meticulous observation over the years to come. Tycho may truly have believed this, for he redoubled his efforts in astronomy after 1572. If Tycho's star did nothing else, his assistant Johannes Kepler later wrote, it produced a great astronomer.

Eighteen years after he had first sighted the new star, in a major work entitled *Astronomiae instauratae progymnasmata* (Preliminaries for the Revolution in Astronomy), Tycho expressed, with a confidence characteristic of the new scientists of his time, just how much the nova of 1572 had altered his thinking:

> I now no longer approve of the reality of those spheres the existence of which I had previously admitted, relying on the authority of the ancients rather than driven by the truth of the matter itself. At present I am certain that there are no solid spheres in heaven, no matter if these are believed to make the stars revolve or to be carried about by them.

If others did not share Tycho's utter rejection of Aristotle, they were at least unsettled by his unequivocal demonstration that new things could happen in the heavens. Tycho's nova became a fact to be reckoned with, a sticking point in the already troubled Aristotelian account of the universe. Thoughtful people began to sense that the long-accepted cosmology was losing its appeal. That was a frightening prospect. What would take its place? An infinite universe in which human existence was inconsequential and life was without meaning?

To make matters worse, another nova appeared in 1600. It was, in retrospect, probably a common nova, for it was far fainter than Tycho's, and it did not last for long. Astronomers gave it little attention; most of the public never noticed it. Four years later, however, just 32 years after the star in Cassiopeia, another brilliant nova (a supernova by modern reckoning) lit up the heavens. It attracted widespread public interest. Tycho was then three years dead, but Kepler, his former assistant, was eminently prepared to receive it.

THE NEW STAR OF 1604

In a sardonic introduction to his pamphlet on the nova of 1604, Johannes Kepler, then Imperial Mathematician to the Holy Roman Emperor Rudolph in Prague, noted the upset caused when Tycho's nova shook the foundations of the Aristotelian world view. The public, he felt, regarded the star of 1572 as a "secret hostile inruption," "an enemy storming a town and breaking into the market-place before the citizens are aware of his approach." The nova of 1604, in contrast, came at a time of great astrological promise, and it would be welcomed by all as a "spectacle of a public triumph, or the entry of a mighty potentate. . . ."

Kepler was referring not to any widespread acceptance of Copernicanism—that was still decades away—but to the fact that the new star of 1604 appeared in a region of the sky, between the constellations of Ophiuchus and Sagittarius, that was already under intense scrutiny by astronomers and astrologers. Jupiter and Saturn were passing close by one another in the sky (an event called a conjunction) and Mars was moving to join the pair. Such conjunctions are rare, but they repeat cyclically in a predictable fashion. The alignment of 1604 was, in the lore of the time, the beginning of an 800-year cycle. The last conjunction of Jupiter, Saturn, and Mars had marked the rise of Charlemagne, and the one before that the Nativity. What would the next one bring?

Astrology and astronomy were not distinct subjects at the time—both Kepler and Brahe cast horoscopes for their patrons—and Kepler was therefore watching the close dance of Jupiter, Saturn, and Mars with considerable professional interest. On the 9th of October he noted the nearest approach of Mars to Jupiter, which formed a bright pair in the constellation of Sagittarius.

The very next night John Brunowski, an official of the Prague court and an amateur astronomer, stepped out to watch the progress of the conjunction of Jupiter and Mars. He was astonished to see, through a gap in the clouds, not two objects, but three. Kepler, hearing the news from Brunowski, remained skeptical. The clouds were thick, Brunowski was agitated, and new stars were supposed to be rare, if not nonexistent. The bad weather continued. Not until a week later, on the 17th of October, could Kepler confirm the sighting for himself. It was fainter than the star of 1572, no brighter than Jupiter,

FIGURE 17: Johannes Kepler. (Yerkes Observatory.)

but it was to remain visible (except for a brief period when the sun was in that part of the heavens) until the following October.

Like Tycho, Kepler was not the first to spy the star that bears his name. A physician in the Italian backwater of Calabria recorded it the night before Brunowski, and the astronomer Altobelli saw it from Vienna on the evening of the 17th. Chinese astronomers noted it the following evening. Because of the intense European interest in the nearby conjunction, a host of trained astronomers caught the nova almost a week before it reached maximum brightness, and the light curve we can reconstruct from their observations gives us a more complete picture of the development of a supernova than all but a few modern sightings. The European accounts also enable modern radio and X-ray astronomers to locate the remnant of the explosion, which resembles that of Tycho's supernova.

Few astronomers of the time were as well prepared as Kepler to

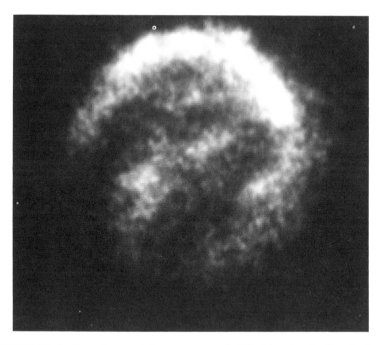

FIGURE 18: An X-ray image of the supernova of 1604, taken by the Einstein X-ray satellite. Its appearance is quite similar to the remnants of Tycho's star and the supernova of 1006. (Courtesy of Fred Seward, Harvard–Smithsonian Center for Astrophysics.)

assess the significance of the 1604 nova. An ardent Copernican and a consummate mathematician, he was at the time preoccupied by the observing records of Tycho. After a fight with his patron, King Christian, Tycho had left Denmark and come to Prague, bringing his journals and notebooks with him. There he entrusted the records to Kepler, hoping that his young assistant would use them to establish the validity of the Tychonic system, that odd hybrid of a stationary Earth and a sun-centered planetary system that the old astronomer regarded as his crowning achievement. Kepler was to find that neither Tycho's nor Copernicus's system precisely fit Tycho's data. The paths of the planets were ellipses, not circles, and they sped up or slowed down depending on how far they were from the sun. The mathematics was lovely, but the results were unexpected, for celestial bodies, in both Aristotle's and Copernicus's system, were supposed to travel in perfect circles at constant speeds. It was left to Isaac Newton, six decades later, to show how Kepler's results could be explained by a new, non-Aristotelian physics.

Because most of his time was spent analyzing Tycho's observing records, Kepler published only a brief report on the nova immediately after it appeared. Two years later he found the time to write a complete book about it. Like Tycho's star, he noted, the 1604 nova remained fixed to the heavens. If it was not as far away as the stars themselves, it was at least much farther away than the moon. But what was it? Like most astronomical writers of the time, Kepler filled most of his pages with astrological speculation on the historical significance of the nova rather than its physical nature, trying to understand what relation it had to the portentous conjunction of Jupiter, Saturn, and Mars. Yet Kepler had little patience with the idle gossip of astrology. He privately regarded most astrology as a "disease," though as the Royal mathematician he felt compelled to speculate himself, and he was ultimately more interested in the relation of the nova to astronomy.

Kepler noted that the new star pulsed and scintillated with a rainbow of colors, an effect we would today attribute to the bending of starlight by Earth's turbulent atmosphere. To Kepler, the twinkling was like the beating of a heart, a pulsation that forced light from the star as blood was forced through the arteries. Regarding the new star as further evidence that change could occur in the regions above the moon, Kepler suggested that the star was celestial in origin, generated when fiery stuff condensed spontaneously from the same material that made the stars. He supported this conjecture with the remark that the stars of 1572 and 1604, as well the fainter nova of 1600, had all occurred close to the luminous glow of the Milky Way. Tycho, in fact, had suggested something similar in his *De Nova Stella*, even pointing out a dark patch in the Milky Way that he supposed was a hole left by the newly formed star. Five years remained until Galileo Galilei trained the first telescope on the heavens, showing that the Milky Way was composed of countless faint and distant stars, not some protean fire. In the absence of such evidence, however, Kepler's explanation of the 1604 nova sounded plausible, if not entirely convincing.

Galileo himself, later to become the most vocal and articulate champion of the Copernican cosmology, was at the time more interested in terrestrial physics than in astronomy. Yet by popular demand, he too took note of the star of 1604, devoting a series of public lectures in Padua to explaining the unexpected apparition. Clearly the public was startled and amazed by the new stars; clearly there was a feeling that a change was coming, that the new stars meant an end to the old established way of looking at the cosmos.

THE NEW COSMOS

Change was indeed in the air, but it was not immediately forthcoming. Two stars and a few skeptical scholars were not enough to bring about a revolution in science. Galileo's use of the telescope, in 1609, played an equally important role. In a few heady weeks of observation the Italian astronomer discovered four moons of Jupiter, circling their planet like a solar system in miniature, clear evidence that other centers of attraction existed beyond Earth. His crude telescope revealed mountains on the moon and spots on the sun, additional blemishes on the supposedly immaculate and immutable heavenly spheres. Later that year, Galileo watched in astonishment for several months as Venus displayed a full cycle of phases like the moon, a phenomenon that was geometrically impossible in an earth-centered universe.

Even these developments did not mark the passing of the Aristotelian cosmos. There were too many issues to be settled first, religious and philosophical as well as scientific in nature. Three decades later, Galileo himself was seized by the Inquisition, taken to Rome, and forced to recant his teaching of Copernicanism. But the die was cast. As telescopic observation became more common, as measurements of star positions became more accurate, informed opinion could no longer favor the old dispensation. The novae of 1572 and 1604 therefore marked not the last gasps of the Aristotelian system, but the opening shots of a revolution.

Poets, sensitive to the spirit of the times, reflected on the meaning of the stars long before their true physical nature was resolved. Abraham Cowley, in an "Ode to the Royal Society," published in the mid-1600s, expressed the general consternation at the passing of the old order:

> So when by various Turns of the celestial Dance
>   In many thousand Years,
>   A Star, so long unknown, appears,

Though Heaven it self more beauteous by it grow,
It troubles and alarms the World below,
Does to the Wise a star, to Fools a Meteor show.

And John Donne, of all the poets of his age, most stricken by the implications of the new astronomy, mused eloquently that things were falling apart. In a poem written in 1611, commemorating the death of a young girl, Elizabeth Drury, the daughter of his patron, he wrote:

And new Philosophy calls all in doubt,
The Element of fire is quite put out;
The Sun is lost, and th' earth, and no mean wit
Can well direct him where to look for it.
And freely men confesse that this world's spent
When in the Planets, and the Firmament
They seeke so many new; then see that this
Is crumbled out againe to his Atomies.
'Tis all in peeces, all cohaerence gone;
All just supply, and all Relation.

Yet there was no turning back. What would Donne have thought of the vast, impersonal universe revealed by the next three centuries of observation? The old Earth-centered world would vanish forever; an infinite vista of stars and galaxies would open; and humanity would no longer occupy a central place in creation. In years to come neither the flaring of another nova nor the passage of a comet would affect humanity as deeply or as indelibly as the two bright stars of Tycho and Kepler.

CHAPTER 5

The Long Wait

> I seem to have stood a long time and
> watched the stars pass.
> They also shall perish, I believe.
> —Robinson Jeffers, *Margrave*

Just as the age of telescopic astronomy began, sightings of new stars ceased. Perhaps mere chance was to blame. Though there's a fifty–fifty chance of getting heads on a coin toss, you can still expect to flip a dozen consecutive tails now and then. It's also possible that astronomers, caught up in the rush of discovery produced by their new telescopes, were occupied with other things. Rather than patrolling the skies, they may have focused their attention on specific tasks: observations of the sun and planets, the discovery of comets, the mapping of the faint stars.

Whatever the reason, from 1604 through the first half of the 19th century, not a single nova (of any sort) was reported. During those lean years, the memory of Tycho's and Kepler's star remained bright. With Aristotle discredited, the notion of change in the heavens seemed less and less outlandish. Observers began to notice stars that fluctuated in brightness from time to time, stars that astronomers today call "variable stars." Armed with a growing knowledge of the different types of changes that stars could undergo, astronomers of the 18th and 19th centuries came to regard the novae of the past as just a particularly violent and spectacular type of variable star.

When sightings of novae became common again, in the late 1800s, astronomers were developing new tools and new ideas. They were exploring the application of spectroscopy and photography to the analysis of starlight. As the study of novae was reestablished and the stock of observations grew, astronomers learned that some novae were far more violent and spectacular than others. In the 20th century a new term—supernova—entered the vocabulary of astronomy. The most spectacular novae of the past, along with a host of new discoveries, began to receive the attention they deserved.

THE UNSTEADY STARS

Enlightenment came slowly. The first variable star appears in the records, unrecognized and unheralded, shortly before Kepler's nova of 1604. In August 1596, David Fabricius, a clergyman, amateur astronomer, and close friend of Tycho Brahe, noted a new star of third magnitude in the southern constellation of Cetus the Whale. He could not find it in any catalog. Was this star, though nearly a thousand times fainter, a distant cousin of Tycho's nova? That seemed reasonable, for like Tycho's nova of 1572, Fabricius's star faded and disappeared after several months. But seven years later the German astronomer Johann Bayer, compiling an atlas of the heavens, catalogued the star again, giving it the name Omicron Ceti.

Bayer was quite unaware, it seems, that Omicron Ceti had enjoyed a previous incarnation as a nova. The star remained an unremarkable entry in his catalog until 1638, when the Dutch astronomer Phocylides Holwarda, after lengthy observations, concluded that the star was playing a game of cat and mouse. Omicron Ceti, or Mira ("the wondrous"), as it is now called, went through recurrent ups and downs in luminosity. Though it was visible near maximum brightness, it was undetectable the rest of the time. The fluctuations, although not strictly regular, repeated themselves about every 11 months.

In 1669, an Italian, Geminiano Montanari, discovered another variable star. This one, to the delight and amazement of observers, went through its paces much more rapidly than Mira. Every 2 days, 21 hours, the brightness of Beta Persei, second brightest star of the constellation of Perseus, decreased sharply, returning a short time later to its original level. Arab astronomers of the Middle Ages had called it Algol, the demon star, and this is its common name today. Did the Arabs know about its strange behavior? If so, they left no direct reference to the fact. Probably the name refers simply to its place among the constellations, where, according to mythology, it represents the head of the Gorgon, Medusa, slain by the hero Perseus.

Though both vary, these first-discovered variable stars do so for quite different reasons. Mira is a giant star, hundreds of times the diameter of our sun. Its outer layers of gas are distended to the point where they cannot maintain their shape and size. Outflowing radiation from the center of the star is absorbed in the outer layers, causing them to heat up and inflate like a hot-air balloon. As they expand, they become more transparent, releasing the trapped radiation, cooling, and deflating back to their original size. Mira thus brightens and dims because its unstable surface swells and shrinks like a huge luminous jellyfish. The light variations we see result from these changes in the surface area and temperature of the star.

It's a different story with the demon star. Shortly after its discovery in 1669 the British astronomer John Goodricke surmised, correctly, that Algol dips in brightness because it is actually two stars, a bright and a faint one, which orbit around each other in slightly less than 3 days. From afar, the two are blended into a single point of light. Every time one passes in front of the other, however, the collective light from the pair is reduced. Such variable stars are called eclipsing binary stars.

Until the advent of photography, during the middle of the 19th century, the discovery of variable stars was excruciatingly slow. Unless a star was conspicuously erratic—like Mira, which varies by a factor of 250 in brightness—astronomers seldom noticed any variability. By 1800 astronomers knew of only a dozen variable stars. But the glass photographic plate, introduced into astronomy in the mid-1800s, preserves starlight in images of silver. With the development of photography, astronomers could compare brightness at one time with brightness at another time, thus discovering tiny variations with relative ease. New photographic surveys swelled the lists of stars that were known to vary. By 1900, astronomers knew of over a thousand variables.

Of all the variable stars, the novae were the most confounding. They were few in number, and seemed to obey no fixed pattern in brightness or duration. In recent decades, we have recognized that most of these novae were the relatively puny flare-ups we now call common novae; a few, like Tycho's and Kepler's, were the powerful explosions we recognize as supernovae. But before this century, such a distinction was unknown. Even the most fundamental questions were unanswered. How far away was a particular nova? What caused the flare-up in the first place? What happened after the star faded from sight?

The long hiatus in nova sightings ended in 1848, with the appearance of a new star in the constellation of Ophiuchus, not far from Kepler's nova of 1604. The 1848 nova never brightened above fifth magnitude, then faded slowly to oblivion. But astronomers could only guess at the nature of the phenomenon they were witnessing, for the spectroscopic work of Bunsen and Fraunhofer was still a decade away.

When the next nova appeared just a few decades later, astronomical techniques had changed dramatically. A few progressive researchers were beginning to employ the first spectroscopes at the telescope. By examining the pattern of spectral lines from distant objects, astronomers could probe the physics and chemistry of distant stars. In 1866, four days after a bright nova was first reported in the constellation of Corona Borealis (the "northern crown"), Sir William Huggins managed to observe it with his spectroscope. The nova spectrum, he noted, showed the characteristic pattern of bright lines produced by hot hydrogen gas, leading astronomers to conclude that they were seeing some sort of cosmic gas explosion, a large-scale version of what might happen if you went hunting for a gas leak with a match. The star had faded to invisibility after nine days, but a faint variable star, known as T Coronae Borealis, remains detectable with modern telescopes.

Spectroscopic analysis revealed similar behavior for an 1876 nova in Cygnus, which astronomers observed this time for several months. As its spectrum weakened, it began to resemble the bright-lined spectrum of a gaseous nebula, one of the clouds of glowing gas seen scattered here and there around the Milky Way; later, the spectrum took on the appearance of an ordinary star, crossed here and there by dark absorption lines.

Astronomers were not sure what to make of this, but many were inclined to regard novae as a transitory stage in the creation of stars, a sort of "missing link" between the fuzzy nebulae of Messier's catalogue and objects like our sun. "No clear dividing-line can be drawn between stars and nebulae," wrote Agnes Clerke, a British astronomer and popular science writer of the time. The nova in Cygnus, the spectroscope revealed, had appeared to be a cloud of gas, a nebula, before becoming starlike and fading out.

The nebula–star connection was not a completely novel idea. Nearly a century earlier, Pierre-Simon Laplace, the great French astronomer and mathematician, had suggested that stars form, together with planetary systems, when clouds of gas contract under their own weight, heating up in the process. If a contracting cloud was rotating slowly to begin with, it should spin faster and faster, like a skater drawing her arms toward her body, and should flatten out into a thin disk as the spin increased. The densest gas, near the center of the cloud, would form a sun; the remaining material in the disk would form planets. Could not many of the nebulae be such gestating solar systems? And might not a nova be the climactic ignition of the central gas? If so, true to its name, a nova would literally be a newborn star.

Like so many young ideas, this notion of the connection between stars, nebulae, and novae incorporated a good measure of truth. Stars and planets are indeed formed along the lines suggested by Laplace. Novae, however, are not part of the process. Newborn stars actually turn on in a much less spectacular fashion, often hidden from sight in the dense clouds that give them birth.

Today we regard both supernovae and novae as aging stars, not infants, undergoing an explosive "change of life" in their final years. But when Clerke wrote, at the close of the 1800s, astronomers knew little about the physical structure and life cycle of stars; they lumped gaseous nebulae (which are clouds of gas in the Milky Way) together with spiral nebulae (which are distant galaxies); they did not know that there were different types of novae. The upcoming century was to be one of sharpening distinctions, expanding physical knowledge, and a realization that things were seldom what they seem at first.

Premature as Laplace's theory was, the relation of novae and nebulae had a great appeal. It helped explain the changing spectra of the two novae that had been seen with the spectroscope. The spectra resembled gaseous nebulae at first, then stars at a later time, simply because nebulae turned into stars. For the time being, in fact, nature seemed inclined to go along with that game: the next nova made its appearance in the midst of one of the most familiar and well-studied nebulae in the heavens, the Great Nebula in Andromeda. It was, in retrospect, the first supernova seen since the time of Kepler and the first to be studied with the tools of modern science.

A BRIGHT STAR IN ANDROMEDA

If our ancestors had eyes as sensitive as a large astronomical telescope, they might have come to worship the Great Nebula in Andromeda. To a telescopic eye, it would appear as a glowing oval, six times larger than the full moon, an imposing presence in the northern autumn sky. Long-exposure photographs resolve the luminosity into a delicate lacework of bright stars and dust. In reality, the Andromeda nebula is a great spiral galaxy like the Milky Way, located about 2 million light-years from the sun.

Even to the naked eye, the Great Nebula is impressive on dark nights, an indistinct blob just north of the corner of the great square of stars in the constellation of Andromeda. It was listed as number 31 in Messier's catalog of nebulae, and is consequently known to astronomers as M31.

A century ago, the distance and composition of M31 were still in dispute. Many thought that it was a nearby gas cloud; others that it was a small group of stars in the Milky Way; others that it was a separate and distant galaxy. Then, as now, it was a favorite object for casual star gazing. On August 17, 1885, Ludovic Gully, a professor of mathematics in Rouen, France, was testing a new telescope at a local public observatory. Pointing it at the usually indistinct blob of the nebula, he was taken aback at the sight of a star, bright and more sharply defined than he expected, near the center of the oval.

Uncertain whether the new telescope was not operating quite properly, or whether the unexpected star might be a bit of moonlight scattered by the mirrors of the instrument rather than a true image, Gully failed to report his observations. Veteran German astronomer Max Wolf saw the same object a week later and attributed it to moonlight as well. There was a full moon on August 25. For several days before and after this date, moonlight made it difficult to distinguish hazy objects like the Great Nebula from the pervasive glow of the sky.

Thus, history accords the discovery of the first supernova since Kepler (and the first outside our galaxy) to a German astronomer, Dr. Ernst Hartwig, on the staff of the Dorpat Observatory in Estonia. On August 20, 1885, while showing a group of visitors around, he pointed the observatory's large telescope at the Andromeda nebula, remarking that, according to Laplace, such cloudlike objects were stars in the making. As he adjusted the focus, he saw a starlike object in the middle of the expected blur of light from the nebula. Hartwig had

FIGURE 19: The Great Nebula in Andromeda, M31. It is about 2.2 million light-years from the Milky Way, and one of the nearest spiral galaxies. (National Optical Astronomy Observatories.)

been observing Andromeda two weeks before, and, despite the brilliance of the moon, was certain that this was a new star that had not been there before. Here, he announced to the assembled onlookers, was a clear example of a star being born from a nebula, just as Laplace had predicted. Other astronomers, he decided, must be notified.

By this date, astronomers had established a system for rapidly alerting one another to urgent items of celestial news. A Central Astronomical Telegram Office, located in the German city of Kiel, could send out news of any new comet, variable star, or other object that might be discovered, enabling major observatories to begin their studies as soon as possible after a sighting. But Hartwig ran into trouble. Moonlight was flooding the sky, making observation of the new star difficult. In fact, had Hartwig not been familiar with Andromeda, he would not have been able to distinguish the star. To Hartwig's dismay the director of Dorpat, a cautious man, refused to send a telegram to the Central Office without confirmation under darker skies.

Baltic rain and fog settled in, delaying the confirming observations. Hartwig, increasingly impatient, posted notes to Kiel, only to find out later that the letters never arrived; their stamps had been stolen from the envelopes, presumably by postal thieves. A week went by before he at last saw the nova again under a dark sky and the director permitted him to telegraph the news. By that time the nova had been observed by others in more favorable locations throughout Germany, France, and the United States. Their observations enable modern astronomers, notably Gerard de Vaucouleurs of the University of Texas, to determine the characteristics of the nova, and to show that, although no one knew it at the time, it was a supernova of unusually great power.

After Hartwig's telegram reached Kiel, the astronomical community turned its full attention to the star. Observers managed to study it for almost six months until its fading light became indistinguishable from that of the nebula around it. At its brightest, the nova was probably of magnitude 5.9, faintly visible to the naked eye. Through a telescope its orange color at maximum light contrasted sharply with the pearly white of the Andromeda nebula, but as it faded it paled to match its surroundings.

S Andromedae, the official name of the new star, was clearly no ordinary nova. It faded too slowly. And those progressive astronomers who observed it with a spectroscope were baffled by what they saw. In contrast to the spectra of the novae of 1866 and 1876, the nova of 1885 showed only a few broad, unrecognizable, dark and bright lines. It took decades to realize the true nature of the object—it was a supernova, an explosion a million times more powerful than a common nova.

THE RECOGNITION OF SUPERNOVAE

The pace of astrophysical discovery quickened, and astronomy seemed on the brink of recognizing the special nature of supernovae. In fact, had astronomers known that Andromeda was a galaxy 2 million light-years away, there would have been little doubt of their "super" nature. At maximum brightness of 6th magnitude, a nova as far away as a million light-years would be emitting billions of times more energy than the sun, a prodigious amount by any standard.

But in 1885, no one knew the distance of Andromeda, or the distance of any nova. And when, with a little luck and a little ingenuity, astronomers were finally able to measure the distance to a nova, the information turned out to be a red herring, delaying the full understanding of both the distances to galaxies and the distinction between novae and supernovae.

It began on the morning of February 22, 1901, when a new star appeared in the constellation of Perseus. Nova Persei 1901 was a common nova, not a supernova. But it was close enough to reach extraordinary brilliance, rivaling the brightest stars in the heavens at its maximum. By August of that year, astronomers noted that the star was surrounded by a few narrow arcs of light, like rings of smoke drifting concentrically outward from the blast. As the months went by, the luminous cloud grew in size at a remarkable rate. Within two years it was bigger than the full moon, though dimming rapidly.

One might think at first that the expanding cloud was a cloud of gas, sprayed outward from the explosion that produced the nova. Lines in the spectrum of the nova, photographed during the first days of the explosion, showed a Doppler shift indicating that a cloud of gas was indeed expanding away from the blast. The speed of the gas could be measured from the size of the Doppler shift: it was about 1000 kilometers per second. This may seem rapid. Over two years, the gas would have filled a volume a few times larger than our solar system. But distances in space are vast, and for a cloud that size to appear as extensive as the full moon, the cloud would have had to be closer than the nearest known star, Alpha Centauri, 4 light-years away. It was difficult to believe that Nova Persei was that near to us— such a close object would surely have been noted long before it exploded. Therefore, reasoned astronomers, whatever was producing the expanding ring of light, it was not the gas from the explosion

FIGURE 20: Nova Persei 1901. This photograph, taken in 1902, shows a faint arc of light surrounding the nova. The arc is a "light echo," produced by light from the initial blast reflecting off of clouds of interstellar dust in the vicinity.

itself. It was something moving much more rapidly—perhaps as rapidly as the speed of light itself.

Well then, suggested some astronomers, why couldn't the ring be made of nothing but light? If there were thin clouds of gas and dust already surrounding the star before the explosion, they should reflect the flood of light produced by the initial blast. But it would take time for the light to travel from the nova to the clouds, and the farther clouds should be illuminated later than the nearer ones. In a sense, what looked like an expanding cloud was really just an expanding echo of the blast, reverberating outward with the speed of light as dust clouds farther and farther from the explosion were lit up. If this were so, the ring of light should have occupied a diameter of 2 light-years after 2 years' time. Because it appeared as large as the moon, one could apply simple geometry to show that a ring 2 light-years in size would be about 500 light-years from us. This, all agreed, was a more reasonable distance for the nova.

Further observations confirmed initial suspicions. If the ring of light had been made of gaseous debris spewed out by the explosion, it should have showed a bright-line spectrum like a gaseous nebula. But spectra of the ring taken in November 1902 looked identical to those made of Nova Persei a year earlier. Clearly the ring consisted of reflected light from the initial blast. Not until several decades after the ring had disappeared did astronomers detect a new circle of light forming at the site of the 1901 nova. It was the slower-moving debris from the explosion finally spreading into view.

Astronomers were impressed by the cosmic scale of it all. Here we were, 500 light-years distant, watching a pulse of light rippling outward through interstellar clouds at almost 200,000 miles per second. More important, however, was the result: this was the first time the distance to a nova had been measured with any degree of accuracy.

Well and good. Where astronomers went wrong was to assume that both Nova Persei 1901 and S Andromedae were the same type of beast, and that both produced the same amount of light. If that were true, the fact that S Andromedae appeared somewhat fainter (about 6 magnitudes, or 250 times) would mean that it was farther away, on the order of 10,000 light-years from Earth. This distance was well within the confines of the Milky Way. S Andromedae, and the Great Nebula in Andromeda that contained it, would therefore be relatively nearby objects.

But there were others who disagreed. Rather than making S Andromedae intrinsically faint and the spiral nebulae nearby, they preferred to have S Andromedae luminous and the nebulae distant. The implication was, in this case, that Nova Persei and S Andromedae were two different types of outbursts, the former puny in comparison to the latter. By 1920, there were clear indications from other quarters that there were indeed two varieties of nova. Astronomers, scanning photographs of M31, had detected nearly a dozen novae in the nebula, all far fainter than the star of 1885. Yet, depending on their preconceived notions about the distances of the nebulae, astronomers interpreted this in different ways.

Heber D. Curtis, in his famous debate with Harlow Shapley on the distance of the nebulae, argued that S Andromedae was a rare type of nova nearly 10,000 times as powerful as the common novae, and that the faint new stars in the Andromeda nebula were common novae like Nova Persei. This would put the Andromeda nebula far beyond the boundaries of the Milky Way, some 500,000 light-years from the sun. Shapley countered that if S Andromedae was as powerful as Curtis said, it would outshine 100 million stars, making it far more luminous than any known star in the heavens. To Shapley such energetic objects were inconceivable; he preferred to conclude that S Andromedae was a normal, well-behaved nova, and that M31 was a nearby cluster of stars.

The dispute over the brightness of novae thus took on cosmic dimensions. Either S Andromedae was a normal nova in the Milky Way (and the other "novae" in Andromeda were feeble cousins), or the universe extended far beyond the confines of the Milky Way. Had I been an observer back then, I might well have taken the cautious approach. After all, Shapley only required astronomers to accept one novel idea—the notion that most novae in nebulae were feeble. Curtis, on the other hand, required two—first, the notion that some novae were unimaginably powerful; and second, the notion that the universe was far larger than previous estimates.

As discussed in an earlier chapter, the issue was not resolved until later in the 1920s, when Edwin Hubble's work on Cepheid variable stars firmly established the great distance of galaxies like the Great Nebula in Andromeda. Curtis, the radical in the debate, was correct on this point. Clearly the 1885 nova was something special, something far more powerful than Nova Persei. Though both were new stars of a sort, it was a mistake to call both by the same name.

The first to offer a name for the more powerful explosions was a Swedish astronomer, Knut Lundmark. In 1925, writing for a British research journal, he referred to the luminous novae observed in other galaxies as "upper-class" novae, and the more feeble ones as "lower-class" novae. Only a few novae belonged to the former group, including S Andromedae and Z Centauri, a nova that had appeared in an irregular nebula 30 years earlier. The "lower-class" were in the majority.

Lundmark's terminology never caught on, probably because astronomers were not quite ready to accept the distinction between the two types of novae. But by the 1930s, the distance scale of the nebulae was general knowledge, and astronomers had begun to accumulate all sorts of information about the brightnesses and spectra of novae. A distinction between the two classes of novae was inescapable.

At the California Institute of Technology in the early 1930s, astrophysicist Fritz Zwicky began to refer to "super-novae" of exceptional power in lecturing to graduate students, and the terminology spread through the grapevine of the astronomical community. Thus, in 1934, when Zwicky and colleague Walter Baade collaborated on a short report, "On Super-novae," for the National Academy of Sciences in Washington, D.C., the name stuck. This was, after all, the time when Superman first appeared on the American scene, and supermarkets and supersales were not far in the future. The time was ripe for superlatives of all sorts. Dehyphenated for convenience, supernovae entered the literature of astrophysics.

SUPERNOVAE DEFINED

As we look back over the past half-century of astronomy, the 1934 paper of Baade and Zwicky appears remarkably prescient. It not only named a new class of astronomical objects, but also concisely and insightfully defined their characteristics. Though its proposals were met with initial skepticism, the paper suggested a program of supernova research that astronomers have followed for over half a century.

Baade and Zwicky were born and educated in Europe. There the similarity ends. Baade, five years the elder, was born in Germany in 1893, where he studied astronomy and became proficient in the demanding task of measuring stellar positions and spectra. This was doing astronomy "the old-fashioned way," tediously, meticulously— the kind of work that produced cautious, conservative researchers.

Trained in the tradition, Baade was nonetheless fascinated by the growing application of physics to the understanding of the stars, and he wanted to do research at the frontiers of the discipline. The largest and best equipped telescopes in the world were in California, and, drawn by the promise of carrying out astrophysical observations with the best equipment available, Baade joined the staff of Mount Wilson Observatory in 1931. Over the next decades he became one of the most respected observers in the astronomical community, known for his thorough, reliable analysis of data.

Sometime before World War II, Baade applied for American citizenship, but during a move he mislaid his papers and never bothered to complete the process. To the amusement of his colleagues—who knew him as a soft-spoken, sweet-tempered man— the United States

FIGURE 21: Walter Baade. (California Institute of Technology.)

government classified Baade as an enemy alien during World War II. But it was all to the good. Baade managed to serve his internment at the observatory on Mount Wilson. There, taking advantage of war-time blackouts in Los Angeles, he obtained the clearest photographs of galaxies anyone had ever seen. This work, complementing that of Hubble, helped establish our knowledge of the immensity of the universe in which we live.

Whereas Baade was the epitome of the established scientist, Zwicky was the eccentric rebel. He was born in Bulgaria, but spent his childhood in Switzerland. After taking his doctorate at the Federal Institute of Technology in Zurich (Albert Einstein's alma mater), Zwicky came to the United States in 1925 and became a faculty member in physics at the California Institute of Technology. Opinionated and imperious, he tended to pontificate rather than propose, infuriating his colleagues even when they knew he was right. (One fellow scientist, the story goes, suggested that the Bureau of Standards give the name "Zwicky" to a unit of abrasiveness. It had a nice sandpapery sound to it. But for practical purposes, another noted, that would be like measuring shoe sizes in light-years; the "micro-Zwicky" would be quite sufficient.)

As he grew older, Zwicky came to regard himself as a prophet ignored in his own time, bringing the insights of physics and his own pet methodology of discovery (which he called "directed intuition"), to the mass of pedestrian astronomers and the general public. He formulated clever schemes for relieving Los Angeles smog, ranging from a plan to subsidize car-pooling to a law banning autos altogether. In the 1950s he pressed for a program to establish manned bases on the moon. He was cranky and difficult to work with; but he was also an excellent physicist. Zwicky's astronomy, much of it devoted to the study of supernovae, places him in the front ranks of modern astronomy. Had he not been so single-mindedly convinced of the importance of his work, the pioneering research in supernovae might never have been done.

Thus, the 1934 collaboration of Baade and Zwicky was, one might say, made in heaven—a marriage of rigor and passion. They began their paper by reviewing the "very curious puzzle" of the more luminous novae seen in galaxies. Regarding the star of 1885 as typical of this rare breed, they calculated that a single such supernova would emit, in a month's time, as much light as the sun does in 10 million years. How to account for such extraordinary power?

FIGURE 22: Fritz Zwicky. (AIP Niels Bohr Library.)

Before the explosion, the authors reasoned, supernovae were probably stars not too much different from the sun. But the energy of the explosion itself was so great that a sizable fraction of the mass of the star must have been converted to energy (according to Einstein's famous $E = Mc^2$) in the process. Thus, they went on, "it becomes evident that the *phenomenon of a super-nova represents the transition of an ordinary star into a body of considerably smaller mass*" (italics in original).

The energy released by this explosion of matter, Baade and Zwicky calculated, could also account for one of astronomy's more puzzling problems: the origin of cosmic rays. Cosmic rays are not rays at all, but fast-moving subatomic particles and nuclei of atoms that continuously bombard Earth from all directions. Where do they come from? If supernovae occur on the average of once every thousand years in a given galaxy, said Baade and Zwicky, they could spew out enough energetic particles to account for the influx we detect. Fifty years later, the origin of cosmic rays is still not fully understood, but supernovae remain a likely source.

The claim that supernovae resulted from the collapse of a star into something smaller seems remarkably prophetic. After all, it was possible to imagine other sources of energy for a supernova. Some favored the notion that supernovae arose from the chance collision of two stars, for instance. What is more, astronomers at the time had only a rudimentary idea of the changes that went on during the life cycle of a star. Half a century was to pass before observation caught up with speculation, and astronomers came to believe that Baade and Zwicky were, in essence, correct.

Even more remarkable, Baade and Zwicky had visualized correctly what might remain after the explosion. In a concluding paragraph, they wrote: "With all reserve we advance the view that a super-nova represents the transition of an ordinary star into a *neutron star*, consisting mainly of neutrons. Such a star may possess a very small radius and an extremely high density." This was speculation of the highest order.

A Russian physicist, Lev Landau, had proposed the idea of a neutron star just two years earlier. A neutron star was a theorist's dream, a star crushed so small that a ball the size of Manhattan Island could contain the entire mass of the sun. At the time, and for nearly three decades following, most scientists regarded such things as borderline science fiction. Neutron stars, even if they existed, should be so faint they could never be detected, and astronomers, like all scientists, don't like to rely on anything invisible or unmeasurable to explain something incomprehensible. Baade himself probably never took the suggestion seriously, for he never returned to neutron stars in later publications.

In contrast to Baade's cautious approach, Zwicky seized on the more spectacular aspects of their work. His lectures on the subject caught the attention of the popular media, and Zwicky seemed to enjoy the fuss. A cartoon carried by the Associated Press on January 19, 1934 summed up their thesis with prophetic brevity. Entitled "Be Scientific With OL' DOC DABBLE" it read:

> Cosmic rays are caused by exploding stars which burn with a fire equal to 100 million suns and then shrivel from 1/2 million mile diameters to little spheres 14 miles thick, says Prof. Fritz Zwicky, Swiss Physicist.

Be Scientific with OL' DOC DABBLE.

FIGURE 23: The 1934 cartoon setting forth Zwicky's theory of supernovae. (Courtesy of the Associated Press.)

Looking back, forty years later, Zwicky offered the following characteristic comment: "This, in all modesty, I claim to be one of the most concise triple predictions ever made in science." A half century of research, capped by the observations of the brilliant Supernova 1987A, bears him out. At the time, however, it was just one more crazy idea.

THE SEARCH BEGINS

Any schoolchild knows that science is built on a foundation of careful observation. Yet when Baade and Zwicky's paper appeared, there were practically no observations to support their bizarre claims. By 1934, only twenty supernovae had been recorded, all in galaxies outside our own; few of these had been followed photographically and spectroscopically. Most of the brighter supernovae in our own galaxy, except for Tycho's and Kepler's star, were known only as exotic references in Oriental texts. No astronomer had yet examined the records carefully; they were simply evidence that someone, somewhere, had seen something out of the ordinary. When all was said and done, astronomers knew practically nothing about the light curves of supernovae, their spectra, the nature of the stars that preceded them, the nature of the debris they produced, or the character of the stars they left behind.

Informed opinion, therefore, did not favor Baade and Zwicky. Most astronomers were willing to concede that supernovae were indeed distinct from novae; but the nature of the explosion was anyone's guess. The obvious task at hand, as Zwicky saw it, was to convince the Philistines. He would find more supernovae, so that astronomers could study them extensively from first flash to last dying light.

Based on the scarcity of supernova sightings at the time, Zwicky had little chance of success. To catch such a rare event, one had to recheck as many stars as possible as frequently as possible for any changes in brightness. The more stars one monitored, the better. For the utmost efficiency, one could look for supernovae in the multitude of galaxies beyond the Milky Way. Each contained a billion stars or so, and any supernova would be so luminous it would clearly stand out from the pack. Thus, a frequent survey of dozens or hundreds of galaxies might reveal, from time to time, new stars in the midst of surrounding nebulosity.

The rate of occurrence of supernovae was a critical unknown. Watching a trillion stars might not turn up supernovae fast enough if they were sufficiently rare. Zwicky estimated, using the tiny sample he had, that supernovae occurred about once every thousand years in a galaxy. Assuming his estimate was good, monitoring a thousand galaxies should turn up one or two explosions a year. If the supernova rate was much lower than that, however, Zwicky might have to watch for a lifetime or longer.

To most astronomers, it seemed a big gamble. Even if the search paid off, the return seemed too small to justify all that effort. Few were eager to take thousands of photographs of galaxies and to make tedious comparisons of the images, just to produce tables of data on a phenomenon no one understood. Still Zwicky persisted in the quixotic task. Writing years later, he portrayed himself as a voice crying in the wilderness (he referred to skeptical colleagues as "the Babylonians among the staid astronomical fraternity"). It would be fairer to say that most astronomers wisely chose to occupy their time with less audacious projects. A longer list of supernovae might be a nice thing to have, but it hardly seemed central to the study of the stars.

Thus, the first systematic search for supernovae was a low-budget affair. Zwicky purchased a large commercial camera lens, and installed it in a mounting on the roof of the Robinson Astrophysics building at Caltech. As often as possible, between 1934 and 1936, he photographed the rich cluster of galaxies in the constellation of Virgo. The project, wrote Zwicky, "accompanied by the hilarious laughter of most professional astronomers and my colleagues at Caltech," was a disappointment. During the entire two years, not a single supernova was detected, though Zwicky had expected to find two or three. The estimate of the supernova rate, it seemed, had been overly optimistic.

While the photographic search went on, Zwicky was already pressing on with plans for a more substantial hunt, using a new type of telescope that had recently been invented by an Estonian optician, Bernhard Schmidt. Most telescopes see only a small portion of the sky at any one time, making it necessary to take a large number of photographs to survey any sizeable area. Schmidt's design, now a mainstay of astronomical technology, photographed wide patches of the heavens with unprecedented clarity and power. Zwicky, determined to have such a telescope for the survey, successfully persuaded his chairman, Robert Millikan, to support the project.

But how to pay for the telescope? George Ellery Hale, one of the country's foremost astronomers, was then involved in the project to build a 200-inch-diameter telescope, the world's largest, on Palomar Mountain. With Hale's help, the Rockefeller Foundation, patrons of the 200-inch project, agreed to allocate $25,000 of its funds to build a Schmidt telescope 18 inches in diameter. Hale and Zwicky justified the small telescope as a complement to the much larger and more costly instrument. While the Schmidt telescope scouted broad swaths of sky for interesting objects, the 200-inch would study much smaller areas in greater detail. From the start, however, Zwicky had the wide-field telescope earmarked as a hunter of supernovae.

On September 5, 1936, the new Schmidt began photographing the sky. By March of the following year, Zwicky had detected his first supernova. In August, he found a second. This one, in the galaxy IC 4182, remains one of the most luminous supernovae ever detected. On September 9, he found a third. Zwicky's colleagues began to take notice. He and Baade followed the progress of the newly discovered supernovae, publishing light curves of the fading stars, while colleague Rudolph Minkowski observed their spectra with the Mount Wilson 100-inch telescope. Within five years, when the World War curtailed observations, Zwicky and his assistant Dr. J. J. Johnson had almost doubled the number of known supernovae. The prospects for future discoveries seemed good—supernovae seemed to be going off every few hundred years in a typical galaxy, faster than Zwicky had anticipated.

Zwicky's success increased professional interest in the study of supernovae. Equally important, perhaps, was the growing suspicion that supernovae played an important role in the lives of stars. Astronomers were beginning to understand the nuclear reactions that produced starlight. They were beginning to realize that stars lived only a finite length of time and that, in their later stages, they might explode with catastrophic consequences. Stellar explosions began to seem more like a stage of stellar life rather than a weird pathology. In short, supernova research was becoming an accepted branch of astrophysics.

As with most research, new discoveries raised new questions. Zwicky's initial survey, bolstered by the spectroscopy of Minkowski, established the fact that there were at least two principal types of supernova explosions. Known today as Type I and Type II supernovae, they differ from one another in luminosity, in the shape of the light curve (a graph of brightness versus time), and in the appearance of their spectra. Why the difference? Even forty supernovae were not enough to draw any profound conclusions. Most of the discoveries were faint flickers in distant galaxies. It was difficult to monitor them for long—especially as the tedious task of comparing photographs delayed the actual recognition of a supernova, sometimes for weeks or months. Those that were caught in time were often too faint for

FIGURE 24: Two views of the galaxy NGC 5253, photographed with the 48-inch Schmidt telescope at Palomar Observatory. The second shows a supernova, SN 1972E, discovered by Charles Kowal while continuing the supernova search begun by Fritz Zwicky. (Palomar Observatory photograph.)

effective spectrographic observations. Zwicky realized he needed a larger and more effective search.

Though other astronomers serendipitously discovered super-novae from time to time, Zwicky did not resume his project in earnest until the late 1950s, when he was able to gain access to a new 48-inch Schmidt telescope on Palomar Mountain. The larger telescope proved enormously successful, and Zwicky increased his own discovery rate to about a dozen new supernovae a year. He also organized teams of astronomers at other observatories to devote telescopes and labor to the project. (Some of these projects continue to this day.) By the time of his death in 1974, Zwicky had been credited with discovering more than 120 of the more than 400 known supernovae; an assistant, Charles Kowal, who worked on the project for another decade, racked up a career record of 80. As of the writing of this book, the list

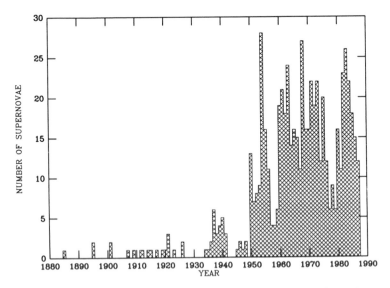

FIGURE 25: A graph showing the number of supernovae discovered each year between 1880 and 1987. The effects of Zwicky's supernova surveys show up as increases in discovery rate. The first was in the late 1930s, when the 18-inch Schmidt went into operation. The second is in the late 1940s when the 48-inch Schmidt was completed. The larger Schmidt was devoted to the task of mapping the entire sky for nearly a decade, and Zwicky did not have access to it until the late 1950s, when the rate rose still further; but there were serendipitous discoveries all the time.

of known supernovae includes more than 650 objects. The accompanying graph shows how the number of known supernovae has increased over the years. The effects of Zwicky's diligence are clearly evident as spurts in the discovery rate during his lifetime and a decline in the years immediately following his death.

HIGH TECH ENTERS THE PICTURE

For all its success, searches like Zwicky's consumed an enormous amount of time and effort. Taking survey photographs is a laborious process; examining the images is tedious and fatiguing. The growing list of supernovae conceals the fact that only scanty information is available about most of them. Supernova hunters sometimes won't recognize a supernova on a photograph until several months or several years after its appearance, too late to study its rate of decline or the nature of its spectrum.

As early as the 1930s, Zwicky had toyed with the idea of utilizing television to analyze the light from his telescope. Technology could not match his requirements at the time: television cameras were not sensitive enough, and observers still had to watch the screens as carefully as they had scanned conventional photographs.

In the last twenty years, a number of other researchers have gone a step further, attempting to improve the effectiveness of supernova discovery by employing computers to speed up the running of the telescope and the comparison of data. Ideally, an automated "supernova early-warning system" could respond quickly enough to alert astronomers to the first brightening glimmer of a supernova.

It should work something like this: Controlled by a computer, a telescope at a remote mountain site senses whether the night is clear enough for observations. If it is, the dome opens automatically and the telescope moves to point at one of several hundred preselected galaxies. It exposes a sensor, nowadays a CCD chip (a highly sensitive solid-state TV camera), and in a short time an image of the galaxy is captured and sent to computer memory. There it is automatically compared with an earlier image of the same galaxy. If the new image matches the old, the telescope moves on to the next galaxy in its program. But if there is a new star in the present image, the computer sounds the alarm, relaying a message to an astronomer's computer terminal: "Look at spiral galaxy NGC so-and-so to confirm presence of a suspected supernova."

The power of such a technique might well justify its complexity. Not only could supernovae be observed as soon as they were discovered, but the rate of discovery could be much greater. An automated telescope, tireless and patient, could easily observe thousands of galaxies a year. Just based on the rate at which Zwicky's search produced supernovae, one might expect a robot searcher to detect a supernova every week, or even every couple of days.

In the 1960s, a first step to an automatic search program was begun by researchers from Northwestern University. Astronomers J. Allen Hynek, William Powers, and Justus Dunlap constructed an automated 24-inch telescope at Corralitos Observatory near Las Cruces, New Mexico. The telescope moved under computer control, but observers sitting in a room nearby watched the images of galaxies on a TV screen, comparing what they saw with photographs of the target galaxies. "We were able to check a new galaxy every minute," recalls Powers, "so that on a long winter night we could check hundreds. Operators worked 2-hour shifts—it was grueling."

During the two years that the Corralitos search was funded, it found 14 supernovae, including a few that had not yet reached maximum intensity. The concept had proved itself, but it involved wearing, mindless labor, which accounts in part for its relatively low rate of discovery. Sterling Colgate, an imaginative and hyperactive astronomer at New Mexico Institute of Mining and Technology, then proposed to fully automate the search, employing computers to find the supernovae as well as run the telescopes. At the time, this was a formidable task. Computers in the 1960s were million-dollar, room-filling monsters, and software available for automation was in its infancy. The computer revolution of the 1970s has brought some changes, but the complex programs for storing and recognizing the changes in galaxy images have proven difficult to write. Colgate's automated telescope, after twenty years of planning, experimentation, and revision, is just beginning to operate, and it has yet to claim a supernova discovery.

But Colgate has been a zealous apostle of supernova searches, and his continued interest in such projects has inspired the most successful current search, run by a group of young astronomers at the University of California at Berkeley. Their project, utilizing a 30-inch telescope at the Leuschner Observatory near the Berkeley campus, is the most fully automated telescope in operation today. As of January 1986, the telescope had accumulated 2000 reference images of galaxies for comparison with the nightly observations. On May 16 of that year, Frank Crawford, an astronomer working on the project, discovered a bright supernova on a new image of galaxy M99. Since then, the group has discovered several additional supernovae.

There is still some human intervention in all these discoveries. Astronomers still must check the computerized images for interlopers. But the Berkeley group has been working diligently on automating the comparison of images. The nightly comparisons, as of this writing, have just been replaced by computer software. Only time will tell whether the Berkeley telescope can produce its goal of a supernova a week.

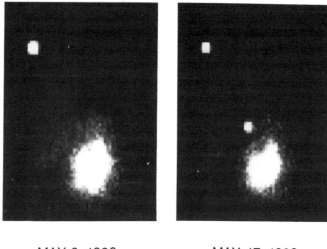

MAY 8, 1986   MAY 17, 1986

FIGURE 26: This is the way the Berkeley automated telescope sees a supernova (this supernova is in the M99 galaxy in the Virgo cluster). On both frames the individual picture elements (pixels) produced by the CCD camera can be seen as small squares. When it is fully operational, the Berkeley telescope will be able to recognize automatically that there is an object on the second frame that is not on the first. (Courtesy of Carl Pennypacker, LBL Automated Supernova Search.)

THE HUMAN EYE VERSUS THE EYE OF THE MACHINE

So far, the automated discovery of supernovae has foundered on the problem of seeing the forest for the trees—computers aren't yet that good at recognizing important features in complex patterns. Whereas the human eye can immediately spot an intruding star in a familiar picture, the computer must labor over the image bit by bit, and it can sometimes mistake a spot of dust or the passage of a cosmic ray for a real supernova.

For the last few years, in fact, the acknowledged champion supernova hunter has been neither a machine nor a professional astronomer, but a 50-year-old amateur, the Reverend Robert Evans of Hazlebrook, New South Wales, Australia. On clear nights, when he is not ministering to his Australian Uniting Church congregation, Evans is likely to be found scanning the galaxies with a telescope set up in his driveway. He averages 20 to 30 hours a month watching the sky—two or three nights' work for an automated telescope. But since 1981, he has managed to discover over a dozen supernovae. No other person has spotted so many using such simple equipment; only three other supernovae, in fact, have ever been discovered by amateurs.

In the late 1950s, as a young amateur observer, Evans became interested in finding supernovae in galaxies. It soon became clear that it would be difficult to avoid mistaking one faint galaxy for another, or to be sure that memory was not playing tricks on him—making

FIGURE 27: Reverend Robert O. Evans holding an award given to him by the American Association of Variable Star Observers in 1983. Since that time he has discovered almost a dozen more, the most recent in January 1988. (Courtesy of Dennis di Cicco.)

him see a change where there really was none. Evans realized that he had little hope of convincing other astronomers he had seen anything unless he kept careful records of the appearance of the galaxies and the surrounding fields of stars. It seemed a daunting task.

Yet with the help of fellow amateurs, Evans eventually prepared a set of charts of the telescopic appearance of hundreds of galaxies. In November 1980 he began to observe in earnest, comparing each galaxy with its charted appearance. On February 24, 1981, he made his first discovery; two weeks later, a second. Now he is a seasoned observer, and professional astronomers have become accustomed to receiving news of supernovae bearing his name.

Evans no longer uses charts as he observes; he simply relies on memory. To Evans, the galaxies are familiar faces—over 1000 fuzzy blobs, each with its own character. During the course of a year, he'll look at hundreds of them. Whenever he thinks he sees an unfamiliar star, he checks an observatory photograph to make sure that the suspected supernova wasn't there all along. If it still seems real, he calls on several other nearby amateurs to confirm the sighting, and then notifies the Central Bureau for Astronomical Telegrams in Cambridge, Massachusetts. Within hours, astronomers worldwide can be monitoring the latest supernova.

To both amateur and professional astronomers, Reverend Evans has become a virtual legend. Like folk hero John Henry, whose hammer put a steam drill to shame, Evans has given computerized telescopes a run for their money. Between 1980 and 1985 he made over 50,000 observations of over 1000 galaxies, and discovered 11 supernovae. Even in the computer age, we have a sneaking admiration for any human who gets the edge on a machine. Evans, the builders of the automated telescopes agree, is the man to beat.

STILL WAITING AFTER ALL THESE YEARS

Ironically, the brightest supernovae appeared long before anyone was prepared to recognize or to study them. Now that many people are watching and waiting, supernovae in our own galaxy seem stubbornly reclusive. Neither Evans nor any of the professional supernova searchers have turned up a supernova as brilliant as the stars of Kepler and Tycho. The brightest modern supernova, SN 1987A, was discovered entirely by accident. (Interestingly, Evans discovered a faint supernova, SN 1987B, the night after the discovery of SN 1987A. It was 10,000 times fainter than SN 1987A; and it was quickly forgotten in the hubbub of the following days.) The supernova searches, after all, are designed to pick up distant events. Spotting a bright supernova requires little help from technology anyway.

Though bright supernovae are rare, faint supernovae seem more frequent than even Zwicky anticipated. Using Evans's observations as a sample, we can simply divide the number of galaxies he observed each year by the number of supernovae he saw in the same period. Of course we must first estimate the number of supernovae he may have missed due to bad weather, as well as those that were drowned in light near the center of the galaxy or hidden by dust within it.

The results are encouraging. In a typical spiral galaxy, supernovae may occur two or three times per century, according to a recent estimate by astronomers Sidney van den Bergh and Robert McClure at Canada's Dominion Astrophysical Observatory. Given this figure, the automated searches of the future may make the discovery of faint supernovae an everyday occurrence.

If the rate of supernovae is as high as one every thirty years, however, the absence of one in our own galaxy for nearly four centuries may seem a bit unusual. But consider this: because of obscuring dust and gas we can only see a small sector of the Milky Way Galaxy that surrounds us. Perhaps many supernovae have gone off, undetected, on the other side of the galaxy, or behind a particularly dense, nearer cloud of dust. If we allow for such unseen explosions, and apply the laws of chance, the current lack of nearby supernovae is no more unusual than a string of five heads in succession.

Still, if the current estimates of supernova rates are to be trusted, we may not have to wait another four centuries for a supernova to rival that of Kepler's. And when the word comes this time—from New Zealand, or California, or from some remote mountaintop—astronomers will know, far better than they did in the days of Kepler, what they are looking at.

CHAPTER 6

Why Stars Explode

> Do not go gentle into that good night,
> Old age should burn and rave at close of
>   day;
> Rage, rage against the dying of the light.
> —Dylan Thomas

SUPERNOVAE OBSERVED

In the sciences, the excitement of discovery is a beginning, not an end in itself. It must be followed by a period of growing distinctions, a period of observation and classification, before new concepts can be established and new questions formulated. As I walk through a museum, I'm astounded by the amount of effort people have devoted to this seemingly secondary task of collecting and classifying. Yet where would science be without it? Fossils, known since antiquity, were not immediately recognized as extinct species—people thought of them as one-of-a-kind monsters, or discarded experiments of the Creator. Only when naturalists had learned to identify and classify a wide range of life forms both living and extinct, could Darwin promulgate an acceptable theory to explain the place of fossils in the history of life.

So it was with supernovae. Back in the early 1930s, when Baade and Zwicky proposed their existence, no one knew much about them except that they were extraordinarily luminous. As the observations accumulated, astronomers began to recognize distinctive features in supernova light curves and spectra. Simplicity became complexity. Eventually, following years of watching and puzzling, things began to come together in a sensible fashion.

Spectroscopy, then, as now, was a main tool of astrophysical analysis. Without a spectrum, one could say little about the physical properties of anything in the heavens. But practically no data on supernova spectra existed when Zwicky began. The earliest supernova spectra were from S Andromedae, in 1885. Because astronomers at that time could not yet photograph spectra—photographic emulsions were too insensitive to respond to the feeble rainbows of light— there were only a few written descriptions of how the spectrum looked through an eyepiece. The sole photographically recorded spectra were of Z Centauri, an 8th-magnitude supernova in the galaxy NGC 5253 that had been observed serendipitously in 1895 by Harvard astronomer Williamina Fleming. In general, spectra of Z Centauri matched the descriptions of the 1885 observers of S Andromedae: only a puzzling pattern of broad dark and bright patches could be seen.

In the late 1930s, as Zwicky started to turn up more supernovae, the base of useful data expanded rapidly. When Zwicky announced a new discovery, Baade would turn the large telescopes on Mount Wilson in that direction, making measurements of the supernova's changing brightness night after night until it faded beyond the limits of detection. If the supernova was bright enough (it takes a considerable amount of light to produce a measurable spectrum even with modern equipment), another colleague, Rudolph Minkowski, would follow its progress spectroscopically. From their work, it soon became clear that there were two principal types of supernovae, which we now call Type I (or SN I) and Type II (or SN II).

Type I supernovae looked like a set of biological clones: each one behaved like all the rest. But Type I behavior, though uniform, was uniformly baffling. After an initial flare to visibility, light from a Type I would rise quickly to a maximum over about two weeks, then fall off at about the same rate for another two weeks. Following this, the rate of dimming slowed, with the light intensity decreasing in what is called an "exponential" fashion. Over a fixed interval of days, light intensity would decline to half its initial level; in the same number of additional days, it would be halved again, and so on until the supernova was no longer visible. The fixed interval for the light from a Type I supernova to drop by half, called its "half-life," was about 50 days.

This exponential behavior was rather suggestive. Radioactive ele-

FIGURE 28: Rudolph Minkowski, who, along with Baade and Zwicky, pioneered the study of supernovae and supernova remnants. (Photograph by Dorothy Davis Locanth. Courtesy of AIP Niels Bohr Library.)

ments, like radium and uranium, give off radiation that weakens exponentially just like the light from Type I supernovae. The half-lives are often longer—uranium has a half-life of over a billion years—but shorter-lived radioactive elements can be produced in the laboratory. Could it be that Type I supernovae were somehow powered by radioactivity? This was a trail later theorists have followed profitably.

The light curves of Type I supernovae were suggestive, perhaps, but their spectra, usually observed while the supernova was at maximum brightness, were a complete enigma. The spectrum was composed not of sharp bright or dark lines like the spectra of nebulae and stars, but of broad bands of light alternating with broad bands of dark. Scientists were not clear whether they were seeing bright light

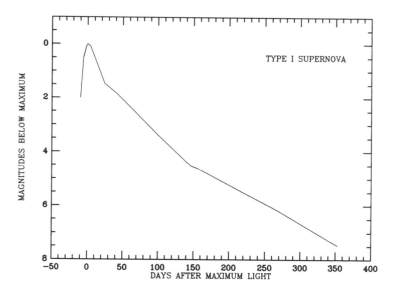

FIGURE 29: Visual light curve of a Type I supernova.

on a dark background or dark absorption on a bright background. For many years, no one could recognize the signature of a single familiar element in this mysterious pattern.

The only thing one could safely conclude was that the broad bands were evidence of the enormous turbulence of the supernova explosion. Spectral lines arise from the collective effect of huge numbers of atoms and molecules in a gas. Each individual atom or molecule absorbs or emits light at a specific wavelength. If all the atoms are at rest, therefore, one sees a spectral line only at that wavelength; it will be sharp and well defined. But if the atoms or molecules are in violent motion, some of them, at any instant, will be approaching and some will be traveling away. Light from the particles traveling toward the observer will be blueshifted, whereas light from the particles traveling away will be redshifted, due to the Doppler effect discussed earlier. The result is a broadened spectral line.

How broad the line is depends on how much motion there is in the gas. The more violent the motion, the greater the range of velocities in the gas, and the broader the resulting spectral line will be. In Type I supernovae, the broad bands indicated speeds on the order of 10,000 kilometers per second. This is a pace seldom encountered in nature. The average molecule of air in the room you are sitting in is traveling less than 1 kilometer per second, and you'd have to raise the thermostat to about a billion degrees to get air molecules traveling as fast as the ejected gas seen in Type I spectra.

The first 12 discoveries made with Zwicky's Schmidt telescope, including the brilliant supernova in IC 4182, were all of Type I. "We almost concluded that there was only one type as far as light curves and spectra were concerned," he wrote (and in fact, Zwicky found nothing but Type I's until his 36th discovery). But number 13 was lucky. Discovered by Zwicky's assistant, Dr. J. J. Johnson, in 1940, it differed significantly from the first dozen supernovae, both in the shape of its light curves and in the appearance of its spectrum.

As more of these Type II supernovae were discovered, it was clear that they were a more heterogeneous group. The light curves and spectra were not always close replicas of each other. But there were recurring characteristics. For one thing, Type II supernovae at maximum gave off 3 to 10 times less light than the Type I blasts. This made them harder to spot; all else being equal, a Type I supernova was visible over a larger distance, and we should expect to see more of them as a result.

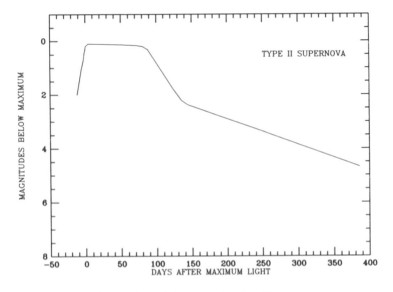

FIGURE 30: Visual light curve of a Type II supernova.

A comparison of light curves revealed that the Type II supernovae spent more time at maximum and declined more slowly than
their Type I cousins. Sometimes the final decline of a Type II appeared to be a well-behaved exponential, but there could be inexplica-
ble jiggles and bumps in the light curve. Based on these irregularities,
Zwicky tried to establish additional classifications of types III, IV, and
V, but the distinction never gained widespread acceptance. Type II
supernovae today include all supernovae that are not Type I.

Like the light curves, spectra of Type II supernovae were significantly different from those of Type I. The most striking difference was
the presence of broad emission lines from hydrogen gas, something
one never sees in Type I spectra. Hydrogen is the most abundant
element in the universe; 9 of every 10 atoms in an average star are
hydrogen. So astronomers were not surprised to find it in the ejected
material from exploding stars. Lines from elements heavier than hydrogen were visible as well, indicating helium, magnesium, silicon,
and others. Because the lines in Type II spectra weren't quite as broad
as those in Type I, it was easier to make the identifications. Apparently the gas in Type II explosions, though expelled with speeds of
thousands of kilometers per second, was a bit slower moving than
that in the Type I explosions.

There was one other striking difference between the two types of
supernovae: they occurred in different places. Type I supernovae,
almost without exception, appeared in the round or oval-shaped galaxies that Edwin Hubble called elliptical galaxies. Type II supernovae,
on the other hand, appeared primarily in the arms of spiral galaxies,
among groups of bright, massive stars, glowing billows of hydrogen
gas, and dark concentrations of dust.

While photographing galaxies during his World War II internment, Baade discovered that these two types of galaxies represent
distinctly different populations of stars. Elliptical galaxies were like a
Florida retirement community—the stars were a geriatric bunch. It
was easy to understand why. There was little or no gas and dust in
ellipticals. New stars are formed from gas and dust, and without
these parent materials one would expect to find few youngsters running around among the aging stars of an elliptical galaxy. One didn't.
Elliptical galaxies were composed exclusively of ancient stars, formed
early in the history of the universe.

Spirals, on the other hand, had a young population, rather like
Ft. Lauderdale during spring break. They were rich in gas and dust,

and star formation was still going on within them. The youngest, most brilliant types of stars were found in abundance in their spiral arms. Whatever distinguished Type I from Type II supernovae, it had something to do with the difference between the aging population of elliptical galaxies, and the younger, more active population of spiral galaxies.

Since the late 1930s, when the distinction between Type I and Type II supernovae first appeared in the literature, astronomers have observed over 600 supernovae. Although many details have been added to the story, the fundamental distinction between the types is still recognized. During the same period, astronomers made major advances in understanding the lives of stars. Physicist Hans Bethe, for instance, discovered the chain of nuclear reactions that causes stars to shine. William Fowler later found out how these nuclear reactions produce all the chemical elements heavier than hydrogen. At the University of Chicago, a brilliant young physicist, Subrahmanyan Chandrasekhar, calculated what stars might look like after their stores of energy had burnt out. (Bethe, Fowler, and Chandrasekhar were to win Nobel Prizes for their work.) Other astronomers, using computers, created elaborate programs to simulate the changes that went on as stars grew older. Baade and Zwicky's suggestion that supernovae marked the catastrophic collapse of stars fit in well with this picture.

Yet to this day, Type I supernovae, largely because of the difficulty of interpreting their spectra, remain somewhat problematic. Type II supernovae, despite wide variations in their behavior, are better understood. We shall focus our attention, therefore, on the story of Type II supernovae before turning to Type I's. Type II supernovae, we shall see, are a direct consequence of the dangerous life-styles of the hottest, most massive stars.

THE LIFE OF A STAR

Stars live their lives on the brink of disaster. Gravity, the force that shapes them and sets them alight, constantly threatens their very existence. When gravity eventually prevails, a star may become a supernova. How can this be?

A star begins its life as a cold, tenuous cloud of gas floating in the expanse of interstellar space. Every bit of matter in the cloud feels a gravitational pull toward every other bit of matter. Gravity makes no distinctions. This collective attraction causes the outer regions of the cloud to weigh down heavily on the inside, compressing the inner gas and shrinking the cloud. When a gas is compressed, its temperature rises—a fact you can confirm by feeling how warm a bicycle pump gets as you increase the pressure in the tires. So the denser the cloud becomes, the hotter it gets, and when it has contracted sufficiently it begins to glow. Nearly two centuries ago, Pierre-Simon Laplace proposed a process similar to this for the formation of the solar system.

Once formed, however, a star needs an internal source of heat energy, a stock of fuel, to keep it from contracting further. Without such a source, it could not shine except by draining heat from its interior. The loss of heat would cause its internal pressure to drop and it would not be able to resist the crushing weight of its outer layers of gas. It would continue to collapse, a victim of its own tendency to pull itself together. But if a star can generate energy, if it can keep itself warm inside, the pressure of hot gases can stave off collapse. It is a tricky balance. A century ago the British physicist Lord Kelvin showed that without an internal source of heat our sun would cool and shrink to nothingness within a hundred million years.

The sun is far older than that. For over four and a half billion years it has remained stable in size and roughly constant in luminosity. The secret of its longevity lies locked in the nuclei of atoms. The combining of lighter nuclei to form heavier ones, the same energy source that fuels the hydrogen bomb, can provide the energy to balance gravity in the sun and other stars. At least for a while.

In order to get nuclei to combine, a star must be extremely hot. As a star contracts under its own weight, the hottest parts are nearest the center, and so nuclear energy generation starts there first. In its formative years, a star heats up until temperatures at its core climb to over 10 million degrees Kelvin. (The unit of absolute temperature is named after the 19th century physicist.) The sun is this hot at the core, and so presumably are other stars.

At this extreme temperature, nuclei of hydrogen atoms jiggle furiously to and fro, occasionally colliding and fusing together to form a single nucleus of helium. A helium nucleus, oddly, contains less mass than the total mass of the four hydrogen nuclei that combine to form it. But the excess mass does not disappear; it is released as energy, heating the star and keeping it from further collapse. Astrophysicists call this fusion process "hydrogen burning," but it is not fire in the usual sense. If the sun were simply a giant ball of flame, it would last only a few hundred thousand years. A star is an upscaled nuclear reactor, not a bonfire.

Stars where hydrogen burning provides a steadying central energy source are technically called "main sequence stars." The sun is a familiar example. There, over 700 million tons of hydrogen gas fuse to helium each second, as the star resists gravitational collapse. Despite this prodigious rate of consumption, the sun has plenty of fuel to keep it going. At its current rate, it will not run out of hydrogen fuel for about 5 billion more years. But not all stars are as long-lived as that.

Main sequence stars more massive than the sun burn their fuel far more rapidly. A star of 20 solar masses, for instance, shines at nearly 10,000 times the brilliance of the sun. As a result, even though it has 20 times the fuel supply of the sun, it burns through that fuel in only a few million years. Massive stars are prodigal suns, enjoying a brief but active life before their energy is spent.

Whatever its mass, a star must eventually run out of fuel. The immediate result, however, is not catastrophic; the star readjusts first and finds a new, but temporary, source of rejuvenating energy. It happens this way: A main sequence star is hottest at the center. That's where the hydrogen is burned and the energy produced. But when most of the hydrogen in the core has been turned into helium, energy production slows down, and gravity gains the upper hand once again. The weight of the star compresses the interior, raising temperatures throughout the star, and igniting some of the unburned hydrogen that surrounds the core. Again the star is stabilized, with energy produced in a layer of burning hydrogen surrounding a core of helium "ash" from the star's main-sequence days.

While this is going on, the outermost layers of the star expand to enormous proportions. Its surface cools to a reddish hue. We call such aging stars "red giants," or "red supergiants" depending on their size, but even the latter term is an understatement at best. Betelgeuse, a red supergiant in the constellation of Orion, is so large that if it replaced the sun, Earth would orbit within it. Its surface would stretch out to beyond the orbit of Mars.

Most of a star's life is spent on the main sequence, and a considerably smaller fraction as a red giant, for the fuel in the layer surrounding the core rapidly burns up. Still, gravity does not win out even then. While hydrogen is burning in a shell around the core, the core itself is contracting and heating. When the temperature reaches about 100 million degrees, helium atoms can fuse to form carbon atoms, releasing more energy.

And so it goes. Heavier elements burn in the innermost regions of the star, where the temperature is hottest. Lighter elements burn in regions farther out. When the helium burning is well along, temperatures at the center of the star rise further, fusing carbon atoms to form oxygen. At the same time, helium burning continues full force in a shell around the core, and hydrogen continues to fuse to helium in a still larger shell surrounding them all. (In these more advanced stages of life, a star's surface may heat up significantly, making it appear yellow or blue.)

In the most extreme stages of this process, heavier and heavier elements—neon, magnesium, silicon, and iron—can be built up by fusion reactions in the core of the star, as lighter elements are consumed in surrounding shells. The star develops a structure like an onion, with the heaviest elements at the center and the unburned lighter elements at the surface.

All stars start along this path into old age, driven along by gravity. But not all go the full distance. To trigger a continuing sequence of nuclear reactions, a star needs enough mass to compress its core to ever higher temperatures. Most stars are not massive enough to sustain the most advanced rounds of nuclear reactions. The sun, for instance, will probably never get to the stage of burning carbon in its core. It will reach the red-giant stage and then, unable to heat its core any more, simply give up the battle against gravity, a burned-out mass of carbon. But stars 20 times the mass of the sun can produce higher central temperatures just by virtue of their weight. They can easily ignite carbon and heavier elements in their struggle to avoid collapse. One might therefore think that there would be no end to it—a massive star could continue to glue together nuclei and produce energy forever.

Not so. The ultimate bottleneck is that, for sufficiently heavy nuclei, further fusion ceases to produce energy. Iron-56, an iron nucleus containing 26 protons and 30 neutrons, is the most tightly bound of all the nuclei. If one adds any additional particles onto it, rather than releasing energy, it drains energy from the surroundings.

The most massive stars can go no further. They run out of energy when their cores are made of iron, surrounded by successive skins of lighter and lighter elements.

Thus, whatever the mass of the star, the fire must ultimately die out. With ashes choking the hearth, a star must finally succumb to gravity. Its weight crushes it ever smaller and smaller. And for some stars, catastrophe results.

Only for a few. Most stars are about the mass of the sun, or less. These stars, we believe, die out gently. When they run out of fuel, they inexorably shrink. Before they are crushed to nothingness, however, they reach a stage, called "electron degeneracy," where the electrons surrounding the nuclei of atoms are packed together as closely as nature permits. The material of the star stiffens and, even without the benefit of additional energy, supports its own weight. The star will then be only a fraction of its original size. The sun, for instance, is now nearly a million miles in diameter. When it has burnt itself out to a carbon ash, it will shrink to a size no larger than Earth.

Such stellar cinders, called "white dwarfs," are bizarre objects by earthly standards. White dwarf matter is so highly compressed that a bowling ball filled with the stuff would tip the scales at over a thousand tons—about the weight of a small ocean liner. No terrestrial material is even a millionth as dense as that. But in the heavens it's a different matter. At the beginning of this century, before the nature of white dwarfs was understood, astronomers knew of a faint star, Sirius B (a companion of the bright star Sirius), which met the description. Though it was a puzzle for many years, astronomers could easily calculate that Sirius B was millions of times denser than normal. Now we know of many others. Though faint and hard to spot, white dwarfs appear to be quite numerous in the universe.

White dwarfs are faint because they are small. As they age, they grow even fainter. Because a white dwarf possesses no source of energy besides the heat stored from its younger days, it eventually cools and fades. In a few billion years it becomes a dark, dense cinder floating through space.

In the early 1930s Subrahmanyan Chandrasekhar worked out a detailed theory of white dwarf interiors. Nature, he found, set a limit to their size. No carbon-rich white dwarf could support its weight if it were greater than about 1.4 times more massive than the sun. Chandrasekhar's limit is fundamental to our understanding of what happens to aging stars: dying stars more massive than the limit cannot shrink sedately into white dwarfs. They must make other funeral arrangements.

Only a small percentage of stars are more massive than the Chandrasekhar limit, but given the billions of stars in a typical spiral galaxy, they represent a sizable number. What is important in determining whether a star becomes a white dwarf, however, is not its initial mass when it is a main sequence star, but its final mass when it exhausts its last bit of fuel. Most stars naturally shed some of their outer mass into space during their lifetimes. Late in their lives, we observe, they produce powerful "stellar winds" blowing from their surfaces that gradually strip them of their outer layers of gas. It may take tens of thousands or hundreds of thousands of years, but most stars that begin their lives with less than about 8 times the mass of the sun manage to end their days below the Chandrasekhar limit.

We can actually observe the results of this outflow of gas. Around many aging stars we see spherical shells of incandescent gas, called "planetary nebulae." The name refers to their appearance— through a small telescope they resemble the hazy disk of a planet— but there is no connection between them and Laplace's idea that nebulae are infant planetary systems. Planetary nebulae, rather, are the products of stellar winds, material jettisoned by a star as it reaches the end of its days. Many stars end their lives this way. For a small minority of very massive stars, however, even this last-ditch effort must fail.

What, then, happens to a massive star that finds itself, after burning its last bit of fuel, a bit too pudgy to meet the Chandrasekhar limit? Briefly put, it collapses, and that collapse turns into an explosion of epic proportions.

OVER THE BRINK

Silicon, an atom with 14 protons and 14 neutrons in its nucleus, is formed in massive stars during the very last minutes of their lives. Ironically, tiny chips of silicon, fashioned into microscopic circuits, tell us most of what we know about the process. Even though astronomers never have been and never will be able to peer into a star firsthand, they can calculate the physical conditions and chemical makeup of a stellar interior by using computers. Computer models of stars are therefore the only laboratories we have for performing "experiments" to see how stars age, burn out, and collapse.

In the last twenty years, a number of theoreticians, among them David Arnett of the University of Chicago, Thomas Weaver at Lawrence Livermore Laboratories in California, and Stanford Woosley of the University of California at Santa Cruz, have constructed elaborate computer simulations of the supernova blast. Their "recipes" for supernovae attempt to incorporate all that we know about the physics of matter under the extreme conditions of catastrophic collapse. The details vary with the size and chemical makeup of the star. Rather than present all the variations in the process, it may be best to give a representative example.

Let us follow the death of a star 15 times more massive than the sun. As it passes into the final stages of its productive life, with heavier and heavier elements burning in its core and surrounding shells, the pace of change picks up to a frenzy. The last stages of energy generation, ending in the development of a core of iron, take only a matter of days, whereas its youth as a main sequence star, with hydrogen burning at the center, lasted millions of years. It is as if nature, aware of what is in store, is pulling out all stops to avoid a final confrontation with gravity.

Gravity, however, must win out. Its store of energy soon exhausted, the star consists of a bloated ball of gas with a dense iron core slightly smaller than Earth, but containing a little more mass than the sun, resting at the very center. Around this lies a thin layer of silicon, and around that shells of lighter elements. The entire star may be somewhere between 50 and several hundred times the diameter of the sun. (The exact size of the precursor star is a matter of some uncertainty at present, for we are not certain how much mass can be lost through stellar winds. Nor are we sure whether it looks red or blue from the outside.)

No more energy can be produced by the iron core, yet it must support the immense weight that overlies it. At first this is not difficult. The core has shrunk to the point of degeneracy. It is, in effect, a white dwarf, rigid and unyielding, embedded in the still sputtering remnants of the star. But as the surrounding layer of silicon continues to burn, showering nuclei of iron onto the core, the mass of the core edges toward the Chandrasekhar limit. Collapse begins within minutes.

As the iron core begins to fall inward under the pull of gravity,

Mass in Interior to Level
(in solar masses, $M_\odot$)

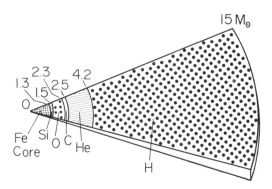

FIGURE 31: The interior of a star 15 times the mass of the sun just prior to a supernova. The iron core (Fe) is surrounded by shells of silicon (Si), oxygen (O), carbon (C), helium (He), and hydrogen (H). (Drafting by Frank Sienkiewicz.)

two things occur to make the collapse even more catastrophic. First, the temperature rises. Ordinarily this would raise the gas pressure and initiate a new round of nuclear burning. But the iron doesn't burn. Instead it begins to break apart. The breakup drains energy from the star, cooling the material just as the melting of ice (which is a sort of breakup of the rigid structure of ice crystals) can cool a pitcher of lemonade. But removing heat from the core of the star only lowers the pressure of the gas, hastening the collapse.

One other event speeds up the catastrophe. As the star becomes more and more compressed, electrons are pushed so close to the protons in the nuclei of atoms that they begin to combine with the protons to become neutrons, releasing a tiny particle called a neutrino in the process. The collapsing core rapidly changes from a dense ball of iron into an even denser glob of neutrons, while the released neutrinos carry energy directly away from the center of the star with high efficiency. This accelerates the collapse even more.

The neutrinos released when electrons combine with protons are odd particles, indeed. They are the closest thing to nothing one can imagine. As far as we can tell, they have no mass, no electric charge, and virtually no effect whatsoever on ordinary matter. Neutrinos can pass with ease through a brick wall, a mountain, or a planet. Under ordinary conditions, a neutrino could penetrate millions of miles of lead as if it were window glass. Once formed in the interior of the collapsing star, an initial burst of neutrinos travels out into space without any difficulty.

But neutrinos carry energy, just as light does. Streaming out of the core of the collapsing star, the neutrinos drain enormous amounts of energy almost instantaneously, in a burst that lasts only a fraction of a second. This rapidly lowers the pressure in the core and removes the last vestiges of an underpinning to the star. In less than a second, the burnt-out iron core tumbles inward, pell-mell toward its own center, where it is compressed to extraordinary densities. All this takes place so fast that the outer layers of the star don't even have time to respond. To a distant observer, the star would seem unruffled for several hours at least, despite the violent events occurring deep inside. Unless one could see the initial burst of neutrinos coming out, the first catastrophic minutes of stellar distress would be completely hidden from view.

What stops the collapse? For some stars, nothing does. If a star is massive enough, it just collapses to a point, becoming an astrophysical oddity called a black hole. Black holes are objects so dense that not even light rays can escape them. Though there is an extensive literature about how they should behave, there are only a few cases where astronomers believe they have actually observed their presence. To discuss the bizarre character of black holes, at this point, would lead us far from the main thread of our story.

In most cases, however, the pell-mell collapse turns into an equally rapid explosion. What causes this, ultimately, is the formation of a neutron star at the center of the star, just as Baade and Zwicky conjectured (though they had no idea of the details of the process in their 1934 papers). When the density of the collapsing core reaches the density of an atomic nucleus—100 trillion times the density of water—the material becomes exceedingly rigid once again. Neutron degeneracy, a situation similar to the electron degeneracy that supports white dwarfs, resists any further compression. The entire core has now shrunk to a tiny ball about the size of Manhattan Island. As quickly as the core collapse began, it ends.

All this happens so fast, however, that material continues to fall inward toward the neutron-rich ball at the center. When it reaches the rigid surface, it bounces. The effect is rather like the shock of impact that reverberates through a sledge handle when one strikes it against a heavy steel wedge. But remember that the neutron core is far denser and more rigid than steel. Almost immediately, layers of stellar material that were falling inward find themselves moving outward again, at speeds of about 10% the speed of light—tens of thousands of kilometers per second.

This speed is far beyond the speed of sound in the gas of the star. Such a supersonic wave of material, like a jet plane traveling past the sound barrier, plows up an expanding sphere of compressed and heated material ahead of it. It is, in the technical language of physics, an expanding shock wave. Similar waves are produced as the heated material from a nuclear blast expands into the atmosphere of Earth. On a much larger scale, the collapsing star has become an exploding bomb—powered by gravity, however, not by nuclear reactions.

Much of this scenario, in fact, was developed initially by researchers working on problems related to nuclear weapons, and many of the initial calculations were carried using programs developed to describe nuclear blasts. Stirling Colgate, a pioneer in the theory of supernova detonation, served as a chief test scientist for the hydrogen bomb program in the late 1950s, and went on to bigger things. In 1966, Colgate and a colleague at Los Alamos National Laboratory, Richard H. White, established the framework of the modern notion of core collapse and rebound.

Since that time, all computer models have had some difficulty in getting the rebounding shock wave to make it all the way to the surface. In the earliest work, the shock tended to peter out before it had produced much of a noticeable effect on the outside of the star. The energetic events at the very core of the star were muffled by the diffuse outer layers, as if one had covered a firecracker with a stack of pillows.

But we know the shock does reach the surface, tearing the star apart; the problem is explaining how. One of the most promising pictures, called "neutrino reheating," invokes those oddball particles, neutrinos, to produce an added lift. The neutrinos come from the neutron-rich core of the star, which, at the instant of collapse, is at extremely high temperatures—hundreds of billions of degrees. At this temperature, the core, for a few brief instants, emits large numbers of fast-moving neutrinos, called, because of their origin, "thermal" neutrinos (to distinguish them from the initial burst of "neutronization" neutrinos produced when the electrons and protons combine to form neutrons). Under most conditions, these neutrinos would simply stream outward from the core through all the outer

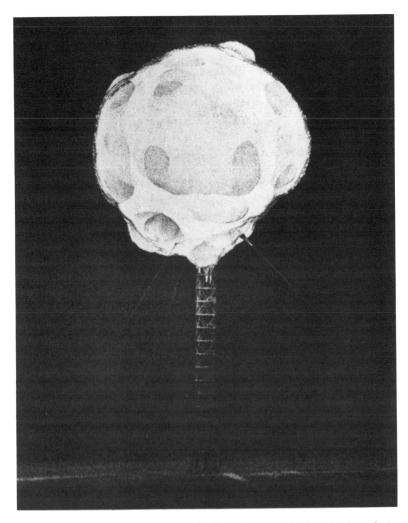

FIGURE 32: A supernova in microcosm: the first microseconds of an atomic explosion. The tower that holds the bomb will be vaporized within seconds. The photograph, recording a test sometime before 1952, required only a 1/100,000,000-of-a-second exposure, and was taken from seven miles away with a lens ten feet long. (Photograph by Harold E. Edgerton.)

layers of the star, into space beyond, but not under the extreme conditions of stellar collapse. Material between the core and expanding shock is by now so compressed that the neutrinos cannot travel through it as if it wasn't there. Some are deflected by atomic nuclei before they have traveled more than a few centimeters.

As far as the neutrinos are concerned, the star offers only a slight resistance—just a bit more than if nothing was there at all. Within several seconds the neutrinos resume their beeline path to the surface, after using only about 1% of their energy to overcome the resistance of the material in the star. However, that 1% of the energy of the neutrinos amounts to 10^{51} ergs, more energy than the sun emits in its entire lifetime. It is more than enough energy to propel the outer layers of the star into space. The star, in effect, feels the friction of the outgoing neutrinos and heats up so rapidly that it blows apart like a cloud of steam from an exploding boiler. Only the dense neutron core at the center remains.

Most of the mass of a star, the equivalent of a dozen stars like the sun, rushes into space at high speeds, from 10,000 to 40,000 kilometers per second. The supernova is now a blazing ball of light. Eons later, when that light has reached our planet, we see it and take note. Vast distances have reduced the incomprehensible violence to a pearl of light, silent against the black of the sky. Another supernova enters the record books.

LIGHT CURVES AND SPECTRA

How does this agree with what we actually see when we look at Type II supernovae? The answer is, rather well. First of all, we would expect to find Type II's solely in spiral galaxies. Only there can we find stars massive enough to collapse in a hurry. The precursors of Type II supernovae, stars more massive than 8 solar masses, shine for no more than about ten million years. Galaxies, however, are billions of years old. In elliptical galaxies, where star formation is no longer going on, all the massive stars must long ago have burnt themselves out. Only in spirals, where the process of star formation continues, should there be young, profligate stars burning at a rapid pace. And there, indeed, is where we observe the bulk of Type II explosions.

Though it's clear that Type II supernovae result from massive stars, astronomers have a tough time verifying the precise details of the supernova explosion itself. All the events we have described in these take place within a few minutes. True, the supernova can't be seen until the exploding star has expanded to hundreds of millions of miles in diameter—which may take several hours—but core collapse and rebound takes place in seconds. No one has ever watched a star pass through its mounting days of crisis and the onset of collapse. Supernovae are only observed after they have risen to enormous brilliance.

Except for the enormous blast of neutrinos emitted from a super-nova, nothing is visible from the outside until the blast wave itself reaches the surface. Then the star becomes a rapidly expanding shell of gas, blown at supersonic speeds into the surrounding space. From afar, we can see a changing level of light as the shell expands and cools. And we can observe the spectrum of the ejected gas. Both the general shape of the light curve, and the spectrum of Type II super-novae seem quite consistent with the model we have presented.

When the explosion first breaks through the surface of the star, its surface balloons to gigantic proportions. Within a day the star is an incandescent ball a billion miles in diameter. To a distant observer, the overall intensity of light grows as the area of the ball grows, for it has more surface area to emit light. Thus, the heating and expansion of the star explains the initial rise of light that we see in Type II light curves.

After several days or weeks, the gas thins, cools, and becomes more transparent. As it does so, the luminosity of the supernova begins to drop. The decline tapers off, however, to a slow exponential decrease because radioactive elements formed in the blast, the most abundant being radioactive nickel and cobalt, slowly decay. The radioactive energy released, much of it in the form of X rays and gamma rays, can effectively heat the expanding cloud of debris. A neutron star formed at the center of the collapse, if it is spinning rapidly and emitting radio waves, can also provide additional heat, as we shall see in a later chapter. Thus, the general behavior of the light curve suggests a cloud of ejected material that glows initially from the sheer energy of the blast, and later from radiation fed to it by the radioactive by-products of the explosion.

Additional information comes from the supernova spectrum. When a supernova is near maximum brightness, the overall appear-ance of the spectrum is a continuous rainbow of light marked by conspicuous bright emission lines and a few dark absorption lines.

These features change as the explosion progresses. To an observer, the supernova first appears a brilliant white. Its spectrum is rich in blue and ultraviolet light, typical of gas at tens of thousands of degrees. As time goes on, the supernova appears redder in hue, a phenomenon that Chinese observers and Tycho noted long ago. The spectrum becomes richer in longer wavelength, red light, typical of temperatures of only a few thousand degrees. We can tell the supernova debris is cooling.

We can also tell its composition. The strongest emission lines are from hydrogen, consistent with the suggestion that Type II supernovae blow off the hydrogen-rich outer layers of massive stars. The lines are Doppler-shifted, and from the shifts we can determine that the gas is expanding with speeds of tens of thousands of kilometers per second—just what we would predict from the computer models. We can even estimate the total amount of material in the shell—it amounts to at least several times the mass of the sun.

All these data can be combined to give estimates of the amount of energy produced by a supernova. During its brief rise to brilliance, a Type II supernova will emit about 10^{50} ergs of light. That's about as much energy as the sun gives off during its entire lifetime of 10 billion years. An additional 10^{51} ergs, ten times more energy, is given off as the motion ("kinetic energy") of the expanding cloud of debris. Computer simulations yield the same values.

Interestingly, however, the energy we detect from a supernova blast represents only about 1% of the total energy produced. The other 99% is carried off by the neutrinos, which, until 1987, had never been observed. The spectacular display of light we see from a supernova pales by comparison to the neutrino luminosity. What catches our attention is—like a rainbow during a rainstorm—a minor side-effect at best.

Nevertheless, the observations seem in general agreement with the theory, and by the 1980s astronomers had growing confidence in the core-collapse scenario of Type II supernovae. They wanted more, of course; they wanted incontrovertible evidence. Because supernovae observed outside our galaxy are so faint, and because they are often not discovered until after they have reached maximum brilliance, observational data were fragmentary and incomplete. Astronomers wanted to see a supernova close up; and they wanted to probe it with every tool available, from visible light to neutrinos. The bright supernova of 1987, the first that could be studied from the very instant of core collapse, was a gift from heaven.

TYPE I SUPERNOVAE

Elliptical galaxies, the home of Type I supernovae, ceased forming stars billions of years ago. Though massive stars may have formed back then, all should by now be burnt out, collapsed to white dwarfs, or neutron stars, or black holes. What, then, produces the supernovae we see in these galaxies?

Because we detect no hydrogen gas in the spectra of Type I supernovae, it seems probable that they arise in stars deficient in hydrogen gas. White dwarfs are likely candidates. They consume almost all their hydrogen and much of their helium during the active stages of their lives, leaving a dense ball of heavier elements behind— primarily carbon and oxygen. As white dwarfs are about the most rigid objects imaginable, how can we get them to explode?

The trick is to dump matter on them from the outside. That idea is not as crazy as it seems. About half of the stars we see are actually members of binary systems, pairs of stars orbiting around each other in a manner not unlike Earth and the moon. If one of the two stars in the binary burns its fuel more rapidly than the other, it will contract to the white dwarf stage while the second star is still going strong. Then, as the second star ages, it will become a red giant, increasing to hundreds of times its normal size. The stage is set for a stellar catastrophe.

Imagine a bit of matter in the region between the two stars. It is caught in a gravitational tug-of-war. Whether it falls toward one star or the other depends on how the mass of one star compares to the mass of the other and how close the bit of matter is to one star or the other. Gas in the outer layers of the expanding red giant can get far enough from its parent star and close enough to the white dwarf companion to be captured by the gravity of the white dwarf. It falls onto the surface of the dense, rigid star, increasing the load the inner layers must bear.

As the pressure on the carbon-rich interior of the white dwarf builds up, the temperature of the star eventually becomes high enough to ignite the carbon nuclei. They begin to fuse to form heavier nuclei of oxygen, silicon, nickel, and other elements, thereby releasing the remaining nuclear energy that the star has not yet extracted. Compared with the frantic pace of a Type II supernova, the burning is not explosive—it flares outward through successive layers of the star like the flame on a Fourth of July sparkler. Astronomers call the process a "deflagration," a burning away, to distinguish it from a more rapid "detonation" of the entire star.

But the sudden deflagration is destructive nonetheless. According to computer simulations done by University of Tokyo astrophysicist Ken'ichi Nomoto and his colleagues, the energy is quite sufficient to produce the effects we see. Most of the star is broken up and scattered into space. No dense neutron core is left behind, as in the case of Type II supernovae. From afar we see an initial flare-up as an expanding fireball of material spreads into space, and a later exponential decline as radioactive material (mostly cobalt) produced by the blast adds a final bit of heat to the debris.

Though Type I supernovae appear brighter, and though they represent the complete destruction of a star, the total amount of energy in the blast is lower than that from a Type II. Only 1% of the energy of a Type II is in the form of light and gas motions. All the rest is in neutrinos, but it can amount to a great deal more than the energy of a Type I. Type I, in contrast, do not form dense neutron stars, and do not produce floods of neutrinos. What energy they have is derived mainly from a last destructive spurt of nuclear fusion reactions. They may produce more light, but that's about all there is to them. When all is said and done, the "sound and fury" of a Type I supernova signifies considerably less than the invisible power of a Type II.

GROWING DISTINCTIONS

Generalizations do not endure for long in a rapidly developing field like astrophysics. In recent years astronomers have begun to recognize supernovae that straddle the boundary line between Type I and Type II. These intermediate types have light curves and spectra resembling Type I supernovae, but they appear in spiral galaxies—an environment where Type I's are never seen. Because of their association with star-forming regions, astronomers believe that they may arise from the core collapse of massive stars, rather than the eruptive burning of a white dwarf. J. Craig Wheeler and Robert Harkness of the University of Texas, who have done considerable work on them, call them "Type Ib" supernovae, and the conventional Type I supernovae, "Type Ia." The distinction seems to be gaining general acceptance within the astronomical community.

Our recognition of Type Ib owes much to recent success in decoding the enigmatic broad-banded spectra of supernovae. As techniques for obtaining spectra have become more sophisticated, spectroscopic data on supernovae have improved. So have methods of analysis. By using computers to simulate the absorption of light in a rapidly expanding shell of gas, astronomers can now "create" spectra artificially, and see whether these synthetic spectra match what they actually see at the telescope. They can add an element here, or change a temperature, adjusting the mixture to reproduce the overall appearance of the real spectrum. Most of the features we see in Type I spectra, we realize now, are highly Doppler-shifted absorption lines of heavier elements, among them oxygen, calcium, silicon, magnesium, and sulfur. These are nuclei formed when a carbon-rich white dwarf begins to burn. No great surprises there.

Two supernovae discovered by Reverend Evans, however—1983N, which appeared in the galaxy NGC 5236, and 1984L, which appeared in NGC 991—seem to be rather different. Though they show no traces of hydrogen—which marks them as Type I—they also show no traces of silicon, which is produced in abundance in the deflagration of a white dwarf. A strong absorption line from helium gas, normally absent in Type I supernovae, is present in both. And both supernovae, unlike most Type I's, seem to be located in regions of a galaxy where star formation is currently going on. There are more than half a dozen other supernovae that resemble 1983N and 1984L. Could they all be massive, young stars?

Wheeler and Harkness, among others, believe they are. Type Ib supernovae, they suggest, are stars that began their lives with a mass between 8 and 25 times the mass of the sun. As they aged, however, they lost most of the unburned hydrogen in their outer layers, leaving behind an onion-layered remnant with helium at its surface. When the ultimate collapse to a neutron star occurred, the ejected material was deficient in hydrogen, and the spectrum was easily mistaken for that of a Type Ia supernova.

We have already seen that mass loss is common among aging stars, but such stars could not have gotten rid of all their hydrogen without help from a binary companion. The second star presumably performs the necessary surgery, using its gravitation to lift hydrogen gas from the surface of its partner. Because binary stars are extremely common, it's not too hard to believe this might happen. But it is not yet possible to scrutinize distant supernovae closely enough to see whether a second star remains after a Type Ib blows up. That would be a conclusive test of the theory.

THE STATE OF PRESENT KNOWLEDGE

From the standpoint of an observer, the nature of the supernova explosion itself is one of the most elusive problems in all of astrophysics. Only two or three stars have ever been detected prior to their explosion—and then only by going back over pictures in observatory archives afterwards. No star has ever been followed through the last few weeks of its life; no astronomer has ever sat by its deathbed and monitored its vital signs. And, until 1987, no star had ever been examined during the critical seconds of core collapse and rebound.

What data we have on the 600-plus supernovae so far discovered are scattered and fragmentary. For some, we have complete light curves from several days or weeks before maximum, and detailed records of their spectra. For others, only partial light curves or one or two spectra exist. And for many, nothing is known. There are a number of supernovae recently that have only been detected as bursts of powerful radio waves in distant spiral galaxies.

The nature of a modern observer's dilemma is easy to grasp. Watching supernovae explode is like studying solar eclipses. One has to be in the right place at the right time, and make efficient use of resources in order to get enough data during a short period of visibility. Eclipses, however, occur with monotonous regularity. The detailed appearance of an eclipse can be predicted thousands of years in advance. Supernovae, alas, kick off without warning. One never knows where they will occur or how bright or faint they will appear.

At the present time, our understanding of the process that produces supernovae relies on things we can control: simulations of the collapse and the resulting blast using digital computers. Theoreticians like Arnett, Weaver, and Woosley have spent decades developing elaborate programs that can be run, with infinite variations, to produce just about any type of explosion one can imagine. Weaver and Woosley have one, KEPLER, named after the great supernova observer of the 17th century. Is the luminosity too low to match the latest observation? If so, they can let KEPLER vary the mix of elements, the density of the initial configuration, the total mass of the outer layers . . . until they get the proper value. The process is a bit like that of an artillery gunner trying to get the proper range in a blustery wind.

One might say that the supernova explosion is, in theory, well understood, too well understood. There are just *too* many options to choose from among the varied phenomena we know take place during the supernova blast. Theory, at the present time, is cheap; the sophisticated computer models are already admirably complete, and there are no obstacles, in principle, to making them even more elaborate.

But observations are dear. The more we can observe supernovae, the more we can learn about which models of supernovae are the right ones, which physical processes are important and which aren't. It is clear that supernovae give us an insight into nature pushed to the extreme. For that, they are of interest not merely to astronomers, but to all who are curious about the physics of the world around them. Just how deep an insight we obtain, however, depends strongly on how much fact we have to set limits on our speculation.

So the definitive chapter on exploding stars cannot be written until many more have been observed in the earliest stages of their appearance. That is why astronomers lionize discoverers like Reverend Evans. That is why they look forward to automated systems of discovery and improvements in spectrographic detectors. And that is why the brightest supernova in four centuries has breathed new life into an already lively field of astrophysical research.

CHAPTER 7

Wisps and Tatters

> I gaze upon the beauty of the stars
>   that cover the face of the sky,
> And think of them as a garden of blossoms
> —Moses Ibn Ezra (1070–1138)

SUPERNOVA REMNANTS

A supernova is an episode of destruction, but it is also an act of creation. As an aging star tears itself apart, hot gases stream outward with speeds near that of light, forming an expanding cloud of luminous material, a "supernova remnant." The remnant may remain visible for tens of thousands of years. In addition, some Type II supernovae leave behind a dense neutron star, spinning rapidly, at the very center of the explosion; exceedingly massive stars may leave behind a black hole. We shall focus our attention on supernova remnants in this chapter, leaving neutron stars and black holes for the next.

Whereas astronomers have only a short time to study the blast itself, they can examine its aftermath at their leisure. Over the last four decades researchers have identified the remnants of virtually all the supernovae known from ancient records. They have recognized well over a hundred other remnants, ranging in age from only a few hundred to tens of thousands of years, using radio and optical telescopes along with orbiting X-ray telescopes. In the last decade, they have even been able to detect supernova remnants in other galaxies, including many in the Large Magellanic Cloud.

The earliest research on supernova remnants was devoted to the two brightest ones, the Crab Nebula and Cassiopeia A. Astronomers were fascinated by their intense radiation and their bizarre behavior. They regarded the Crab Nebula as particularly puzzling. In the 1950s and 1960s it received so much attention that astronomer Geoffrey Burbidge quipped, "You can divide astronomy into two parts: the astronomy of the Crab Nebula and the astronomy of everything else." In time, however, the attention paid off. Just as studies of the sun have helped us better understand the nature of stars, so our studies of the Crab Nebula and Cassiopeia A have enriched and deepened our knowledge of remnants far fainter and far more distant.

THE CRAB NEBULA

Dr. John Bevis, a London physician, discovered the Crab Nebula, though he didn't give it its distinctive name. One of the most active observers of the heavens in Georgian England, Bevis maintained a private observatory outside the city and, about the year 1745, compiled a set of ornate star charts, the *Uranographia Brittanica*. It almost didn't see print. Engraving costs were so high that the printer he had hired went bankrupt; creditors sued for the assets, among them the copper plates for Bevis's charts. Only a few sets of proofs, struck before the lawsuit, circulated among astronomers. Bevis, undiscouraged, kept up his observing, and died 26 years later from a fall on the stairs of his observatory.

The same year that Bevis died, Charles Messier published a preliminary edition of his famous list of nebulous objects. Its first entry, Messier 1 (M1) to modern astronomers, listed a diffuse cloud of light in the constellation of Taurus, the Bull. No stars were visible among the nebulosity. In later editions, Messier credited its discovery to Bevis, who had included the positions of 16 known nebulae among the stars in his charts.

What could the nebula be? William Herschel, who observed it not long after Messier, thought he could see individual stars in the cloud. In the mid-1800s, Lord Rosse, using his mammoth 6-foot-diameter telescope, discerned thin filaments of light in the nebula, and surmised that, with more optical power, these would turn out to be groups of stars. In 1844 he published a first sketch of M1, a rough rendering that made it look a bit like a scorpion or a crab (though some see the indistinct shape as more of a pineapple). Rosse called it the Crab Nebula. Though he drew it again, quite differently, in later works, the name "Crab" had an exotic ring, and astronomers adopted it enthusiastically. The Crab it was, and the Crab it remains.

The Crab Nebula actually looks nothing like a Crab. Isaac Roberts, who first photographed it in 1892, remarked that very few of the previous sketches of nebulae resembled their photographic images. Photographs of the Crab show a squat misshapen blob of light, vaguely like a fat letter S, pierced throughout by a network of bright twisted filaments. There were stars visible, but it was clear that virtually all of them were foreground or background objects. Most of the light came from some gaseous substance.

Spectra of the Crab Nebula confirmed this. The spectra showed bright emission lines, characteristic of a cloud of rarefied gas. A star would have produced a continuous spectrum split by dark absorption lines. The bright emission lines in the Crab were later identified as arising from atoms of oxygen, hydrogen, helium, neon, and sulfur, not uncommon in interstellar clouds. Unlike many gaseous nebulae, however, the Crab emits some light at wavelengths between the bright lines—a continuous background of unknown origin. The Crab Nebula appeared unique in this regard, and the mystery of the continuous radiation was not solved until after the advent of radio astronomy in the late 1940s.

Meanwhile, astronomers used the accumulating photographic evidence to conclude that the Crab Nebula was billowing outward like a cloud of steam. In 1921, Mount Wilson astronomer John Duncan compared the positions of filaments in two pictures of the Crab taken 11 years apart. During that interval, he noted, the edge of the Crab had moved outward, just as the cloud of light around Nova Persei had expanded 20 years earlier.

Was the Crab actually expanding, or was this, like the fast-moving halo around Nova Persei, merely an echo of light? (See Chapter 5.) This time spectroscopic evidence favored a real expansion. In 1913, Vesto Slipher had obtained sharp, highly resolved spectra of the Crab that showed each bright emission line to be a pair of distinctly separated lines, one of slightly longer wavelength than the other. In the 1930s, Nicholas U. Mayall pointed out that this was precisely the spectrum one expected from an expanding cloud of gas, as long as the gas was transparent enough so that one could see light from the far side of the cloud. The shorter-wavelength line of each pair was emitted by gas at the nearer side of the nebula, which was traveling toward the observer; the gas motion produced a Doppler shift to shorter wavelengths (blueward). The other line of each pair, Doppler-shifted to longer wavelengths, was emitted by gas at the far side of the nebula, which was receding from the observer. From the separation of the two lines, one could determine the expansion rate of the nebular gas. For the Crab Nebula this turned out to be about 1000 kilometers per second. Gas moving at this speed could produce Duncan's observed expansion of the Crab if the latter were at a distance of about 5000 light-years from us. The Nebula seemed comfortably far away. There was no need to explain it as a light echo.

A number of years passed before anyone made the connection between the expanding Crab Nebula and the Chinese "guest star" of 1054, even though translations and lists of the ancient sightings had already appeared in several publications. In 1928, Edwin Hubble took a moment from his work on the expansion of the universe to write a short popular paper on the expansion of the Crab. To reach its current size at the rate measured by Duncan, he noted, would take about 900 years. The position agreed with that of the star of 1054. The conclusion seemed firm, but few professionals read Hubble's little paper, and there were only a few who had any interest in supernovae.

Fifteen years later, however, the work of Zwicky and Baade was beginning to gain acceptance. Baade, stuck on Mount Wilson during the war, composed several papers identifying the observations of Tycho's star of 1572 and Kepler's star of 1604 with Type I supernovae. At the same time, in occupied Holland, scholar J. J. L. Duyvendak completed an extensive study of the Oriental records, among which he found many eloquent descriptions of the spectacular events of 1054. These accounts, smuggled past German lines and published in an American journal, revived the study of the Crab Nebula. If the ancient records were to be trusted, the 1054 "guest star" was not an ordinary nova. At a distance of 5000 light-years, it would have been as luminous as a billion suns. And if the Crab had indeed been born at the time of the sighting, we had, at last, incontrovertible evidence of the destructive power of supernovae. Though Baade had already looked for remnants of Tycho's and Kepler's supernovae, he had not yet found them. The Crab was the first undeniable relic of a supernova blast.

In the decades after the war, the Crab became a centerpiece for astrophysical research. Astronomers were rapidly developing new ways of looking at the heavens, using radio waves, and X rays. Whatever method they used, the Crab was one of the strongest sources of radiation they detected. And one of the most peculiar.

The first radio telescopes, built in the late 1940s, soon detected a bright radio source at the position of the Crab. (Astronomers referred to it as Taurus A, the first radio source discovered in the constellation of the Bull.) It was hard to understand how an object could be so luminous at both optical and radio wavelengths. The standard laws of radiation from a hot gas didn't predict that that much radio energy should be generated at all. Yet there was the Crab, giving off 100 times more radio energy than the sun emits at all wavelengths put together. Where did the energy come from?

Iosif S. Shklovsky, a Russian astrophysicist, solved the problem in 1951. The radio waves were generated by a process already familiar to nuclear physicists, but virtually unrecognized by astronomers of the time: the contorted motions of fast electrons under the influence of magnetism. Electrons traveling at high speeds will deviate from a straight-line path in the presence of magnetic fields. If you imagine the magnetic field as a set of lines, resembling the pattern you get when you sprinkle iron filings around a magnet, then the electrons travel along paths that coil like helical springs around the lines of force.

Nuclear physicists used this property of electrons in "atom smashers" of the time. In one type of device, called a synchrotron, the electrons (or other electrically charged particles) were steered by magnetic fields while they were accelerated to velocities approaching the speed of light. The high-energy electrons could be used to penetrate and probe the nuclei of atoms. But physicists who used synchrotrons noted that as the electrons were bent by the magnetic field, they emitted high-intensity light and radio waves. You could see the eerie glow whenever the machine was turned on. It was a nuisance to the physicists, for the so-called "synchrotron emission" drained energy from the electrons, and kept them from being accelerated to higher and higher speeds. But to astronomers like Shklovsky, it was the key to a mystery.

Shklovsky proposed that the Crab Nebula was acting like a cosmic synchrotron. If the Crab was filled with electrons traveling near the speed of light, and if a fairly small magnetic field was present, the electrons would spiral around the magnetic lines of force, emitting strong radio waves and some visible light. Though the source of the electrons remained unclear until the late 1960s, when a pulsar (a spinning neutron star) was identified in the heart of the nebula, there was a way to test the theory. If the radio waves and light were generated by synchrotron processes, the radiation from the Crab should be polarized.

Waves are said to be polarized when they vibrate up and down in just one plane. If you imagine light waves to be like waves you get when you tie a string to a doorknob and wiggle the other end, then polarized waves are produced when you move the end of the string up and down in a uniform, straight line. Light from ordinary hot sources, like a neon lamp or an incandescent bulb, is unpolarized— the plane of vibration of the waves jumps from one orientation to another randomly.

Stars emit light that is unpolarized. But sunlight reflected obliquely off a puddle is not; the reflection process produces some polarization. You can test it yourself. The lenses of "Polaroid" sunglasses

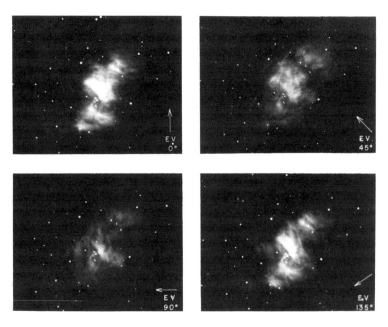

FIGURE 33: The Crab Nebula in polarized light. The four photographs show the nebula through filters that only admit light polarized in one plane (shown by the arrows). The variations in appearance show that the radiation from the Crab is polarized. (Palomar Observatory photograph.)

pass light only of one plane of polarization—they act like the bars of a cage, only permitting waves to pass through if they are vibrating parallel to the bars, not perpendicular. Holding the glasses at arm's length, stand a few yards to one side of a puddle and view the reflection of the sky through one lens while you rotate the sunglasses. You will note that the brightness of the reflection varies. When the plane of polarization passed by the lens is parallel to the plane of polarization of the reflected sunlight, the image is brighter than when the sunglasses are turned at right angles to the plane of polarization.

Astronomers could perform the same experiment with the Crab Nebula. If they put a polarizing filter, like a Polaroid lens, in front of their photographic plates, the appearance of the Crab seemed to change with the orientation of the filter. Radio telescopes could, by reorienting their antennas, measure the same polarization effect. The radiation from the Crab was polarized, just like synchrotron radiation. Shklovsky was right: the Crab was a giant particle accelerator.

The synchrotron mechanism produces only part of what we see from the Crab: the continuous radiation from the featureless, S-shaped blob that underlies the filamentary network of the Crab. In contrast, the complex network of bright tendrils snaking throughout the Crab seems to emit only emission lines from atoms like hydrogen, helium, and oxygen. This light is unpolarized; it is a hot thin gas. The accompanying photographs, both reproduced to the same scale, show the distinction quite clearly. You may recall that early spectra of the Crab showed a series of bright emission lines superimposed on a continuous background of light. Now the meaning of those spectra is clear: the emission lines come from hot gas in the Crab's filaments. The continuous radiation, synchrotron in origin, comes from the main body of the Crab.

Shklovsky's notion of synchrotron radiation solved one problem—the origin of the radio and optical radiation of the Crab—but raised another. As the electrons that produce the radiation give off their energy, they slow down and cease to radiate any longer. To maintain the glow of the Crab, a continuous replenishment of energy is necessary. Where does it come from? The source of synchrotron energy seemed an utter mystery. The mystery was compounded as astronomers looked at the Crab at other wavelengths.

In the early 1960s, astronomers began to observe the skies using X-ray detectors. Gases at temperatures of millions of degrees naturally emit X rays, a form of very-short-wavelength electromagnetic

FIGURE 34: The Crab Nebula photographed through a filter that admits only the continuous light produced by synchrotron radiation. (Courtesy of R. P. Kirshner, W. P. Blair, K. Long, F. Winkler/National Optical Astronomy Observatories.)

radiation, and there are many astronomical objects that should be this hot. But Earth's atmosphere absorbs X rays quite effectively, making it necessary to use rockets to carry X-ray telescopes to heights where they will be effective. In 1962 an X-ray detector, little more sophisticated than a Geiger counter, was flown above the atmosphere by a group of engineers from Cambridge, Massachusetts. It detected a strong source of X rays in the general direction of Taurus. Within two years, other rocket experiments had located the precise position of the source. It was the Crab Nebula, as prominent in X rays as it had been at radio wavelengths.

Yet what powered the X rays? Was it the same mysterious source of energy that maintained the synchrotron radiation? The energy required to produce the light and radio emission from the Crab was enormous, even by astronomical standards; now even more energy was needed to power the X rays. The more astronomers learned about the Crab, the more curious it seemed.

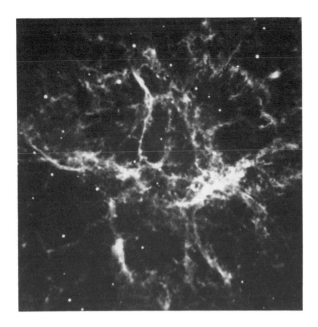

FIGURE 35: The Crab Nebula photographed through a filter that admits only light at 5007 angstroms. This is emission from doubly ionized oxygen (oxygen with two of its electrons stripped away), which is concentrated in the gaseous filaments of the Crab. (Courtesy of R. P. Kirshner, W. P. Blair, K. Long, F. Winkler/National Optical Astronomy Observatories.)

The surprisingly rapid rate of expansion of the Nebula raised additional difficulties. By dividing the size of the Nebula by the observed rate of motion of the filaments, astronomers could calculate when the expansion began. The date was 1140, give or take about 10 years. Walter Baade, the first to do the detailed calculation, assumed that given the errors in the measurements, this was close enough to the 1054 date recorded by the Chinese. In her 1968 Ph.D. thesis, astronomer Virginia Trimble took up the problem again. Her more precise analysis of the nebular expansion left no doubt that the filaments were in fact expanding too fast to have started much earlier than 1140. If that was the date of the supernova that formed the Crab, could the 1054 star be a different object altogether?

Most astronomers found that hard to accept. If there were two supernovae in Taurus within 90 years of each other, then where was the other remnant? And why weren't there Oriental sightings of the second explosion? It was easier to believe that the expansion of the nebula had accelerated over the intervening centuries. Gas that was once moving more slowly had speeded up, making astronomers underestimate how long it had been moving. The nebula, more energetic than it was in the past, looked younger than its years. But a puzzle still remained: the expanding gas cloud should have slowed down, not speeded up, as it plowed into the surrounding interstellar medium (a sparse sprinkling of atoms floating between the stars). To speed up the expansion required, once again, a source of energy, an added push to hurry the nebula along against the resistance of the interstellar gas.

By the late 1960s the Crab Nebula was a threefold puzzle. What kept the spiraling electrons going, maintaining the intense synchrotron radiation at radio and optical wavelengths? What source of energy caused the nebula to emit X rays? What had accelerated the nebula since the explosion of 1054? Nicholas Mayall, writing in 1962, used Winston Churchill's famous phrase to describe it: "A riddle, wrapped in a mystery, inside an enigma." No other supernova remnant seemed quite like it.

There were, however, a few tantalizing clues that pointed to a nondescript star not far from the center of the Crab. If astronomers measured the motion of the wisps and tangles of the Crab and extrapolated back in time, they could determine not only the apparent date of the explosion, but the place where the expansion began in 1054. The position was close to, but not precisely on, a faint, 15th-magnitude star.

Was this the star that blew apart in 1054? If so, why wasn't it located at the precise spot the expanding nebula emerged from? The answer was simple: it was there in 1054, but it had moved since. Astronomers actually measured its speed and direction and traced it back to where it was in 1054: right at the center of the expanding cloud of gas. Baade and Minkowski noted this coincidence in their 1942 study of the Crab, and the star immediately became a focus of attention.

Baade and many later observers noted changes in the appearance of the Crab that seemed to emanate from the central star. Faint wisps of light appeared in the nebula, radiated out through the surrounding gas like ripples on a pond, and disappeared several months later.

Somehow the faint star was stirring up the nebula, creating effects that somehow powered the overall expansion.

This was the state of affairs in the mid-1960s. The key to understanding the Crab was indeed locked in the central star, but astronomers at the time had little idea what it was. The discovery of the source of the Crab's energy came only with the recognition that there was a neutron star at its center, just as Baade and Zwicky would have predicted. That is a story we will relate in the next chapter.

THE SECRET LIFE OF CASSIOPEIA A

In 1942, an engineer named Grote Reber published a map of the sky in a technical astronomy journal. This was no ordinary map: Reber, whose hobby was ham radio, had produced it using a 31-foot-diameter radio dish he had built in his backyard in Wheaton, Illinois, a suburb of Chicago. Instead of showing stars, it showed places in the sky that were emitting strong radio signals. During the several years Reber was conducting his survey, his neighbors, I imagine, must have scratched their heads, for the satellite TV dishes so common today were unheard of at the time, and radar was still a military secret. Astronomers were no more familiar with radio observations than the public, Reber later recalled, for they "could not dream up any rational way by which the radio waves could be generated." They regarded his survey as being "at best a mistake, and at worst a hoax."

Reber's radio map, the first of its kind, showed that most of the signals came from three distinct sources in the heavens. Just a few years later, astronomers identified one source as the center of our own Milky Way; another as a distant galaxy in the constellation of Cygnus, the Swan. In 1948 Martin Ryle and F. Graham Smith at Cambridge, England, using a captured German radar dish, cataloged the third source and gave it its name: Cassiopeia A. Since that time Cas A (as most astronomers call it) has become one of the most studied radio sources in the sky.

From the very outset Cas A presented problems. At radio wavelengths, it was overpowering. If our eyes could see radio waves, it would be the single most conspicuous object in the nighttime sky. But when astronomers looked in its general direction using ordinary optical telescopes they saw nothing. Part of the problem was that radio maps of the time were fuzzy and indistinct. It was difficult to tell the precise location of Cas A. (You have the same problem correlating your sense of sound and sight. It's easy to hear there's a fly somewhere in your bedroom. It's difficult to see where to swat it.) It wasn't until 1951, using an improved technique called radio interferometry, in which several radio telescopes are hooked together to produce more detailed images, that Smith was able to establish a precise position for the source.

At Palomar Observatory in California, Walter Baade and Rudolph Minkowski turned the new 200-inch telescope in the direction given by Smith. There they found the incriminating evidence: a small, roughly circular patch of sky littered with hundreds of faint shreds of gas. Examining the shreds with a spectroscope, Minkowski found they were of two varieties. Some of the gas, rich in oxygen, seemed to be hurtling away from the center of the Cas A region with speeds of several thousand kilometers per second. Mixed among these shreds

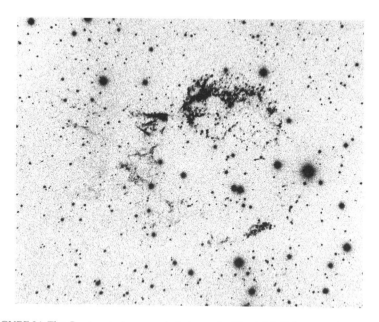

FIGURE 36: The Cassiopeia A remnant photographed in visible light using the 200-inch telescope on Palomar Mountain. (Palomar Observatory photograph by Sidney van den Bergh.)

were others, rich in oxygen, hydrogen, and nitrogen, which did not seem to be moving much at all.

Cas A was evidently a supernova remnant. Only a supernova could have produced the fast-moving knots of gas. By the 1970s, some shreds had moved so noticeably that astronomers Sidney van den Bergh and Karl Kamper were able to measure the expansion by comparing new photographs with Baade and Minkowski's original set. Working backward in time, they found that the explosion that created the remnant occurred sometime between 1653 and 1671. Its distance was approximately 10,000 light-years from us.

The slow-moving shreds of gas suggested a different origin. If they had been expelled from the same center as the faster-moving gas, they must have started several tens of thousands of years ago. Most likely they were bits and pieces of matter shed by the aging star *before* it blew up. Losing mass by "stellar winds" of this sort seems common among the blue and red supergiant stars that mark the last epochs of stellar life. Probably Cas A was just such a star. Later, after it exploded, the shock wave from the blast caught up with the jettisoned debris of earlier days. Thus, the slow-moving gas is a relic of the late life of the star, the fast-moving gas, a product of its violent death.

Because Cas A is such a bright source at radio wavelengths, astronomers have lately been able to examine its structure in great detail. The radio image you see in Figure 38 was produced in 1983 using an advanced radio telescope near Socorro, New Mexico called the VLA—the Very Large Array. The VLA consists of a set of twenty-seven 75-foot-diameter dishes, mounted on movable rail cars, which can be spaced out across the desert over a region 21 miles across. By combining the signals from each detector and using a great deal of computer processing, we can produce images like the one shown. Twenty minutes of computer time on an advanced Cray supercomputer (ordinarily used to create the fancy special effects in movies like *Star Wars*) was needed to produce the image you see. The gas cloud from the supernova appears like a fine shell of material filled with gauzy wisps of gas. In recent years astronomers have been able to watch these wisps change in position and size, something that would have been impossible using the crude radio telescopes of 1948.

Cas A has also been detected by the Einstein X-Ray satellite. There it shows an unremarkable ring structure, just like the remnants of Tycho's and Kepler's supernova. Because it is so young (a third the age of the Crab Nebula), the combined optical, radio, and X-ray data give us vital information on the early development of supernova remnants.

Yet if there was a supernova so close to us just three centuries ago, why wasn't it recorded? The telescope was a common instrument by then. Perhaps the flash was hidden by clouds of interstellar dust—that particular region of Cassiopeia is filled with intervening material, and Cas A is about 10,000 light-years away. If so, the supernova might never have become bright enough to see with the naked eye. Radio waves, however, can penetrate dust clouds, and that is why the remnant remains such a conspicuous object at those wavelengths.

Even if it were hidden by dust, however, a supernova as close as Cas A might have been visible to telescopic observers of the late 17th century. One of the most industrious of these was John Flamsteed (see Chapter 1), the first British Astronomer Royal. During the later decades of the 17th century, he was immersed in his telescopic survey of the sky—which culminated in the definitive catalog *Historia Coelestis Britannica*, published in 1725. The survey required that Flamsteed spend every clear night at the telescope, meticulously measuring the positions of the stars. If there was a supernova to see, Flamsteed might well have seen it.

On August 16, 1680, he recorded a star of magnitude 6 (just barely visible to the naked eye), later including it in his catalog as star number 3 in the constellation of Cassiopeia. By the time the catalog was published, however, Flamsteed had died. Observers could no longer find any star corresponding to the position of star number 3, and so it was dropped from later editions of the publication.

In 1980, William Ashworth, a historian at the University of Missouri, suggested that Flamsteed had indeed seen the supernova that gave birth to Cas A. The position of the supernova remnant is almost the same as that of Flamsteed's star number 3. Only the date seems wrong. From the speed of the fast-moving knots of gas, the supernova should have exploded about twenty years earlier, according to Kamper and van den Bergh. What could account for such a discrepancy?

Suppose the expanding remnant had slowed down over the 300 years since the supernova went off. Then the measurements of its expansion today would indicate that the gas had been moving for longer than it really had (just the opposite of what we saw in the case

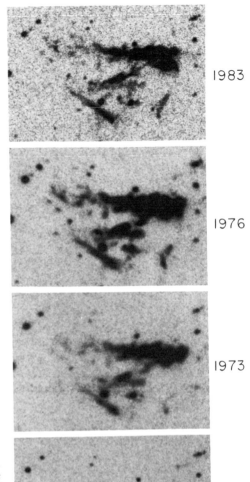

1983

1976

1973

1958

FIGURE 37: A series of photographs illustrating the motion of fast-moving gas in Cassiopeia A. The knots of gas pictured here are found at the upper left of the previous photograph of the entire remnant. (Palomar Observatory photographs by Sidney van den Bergh.)

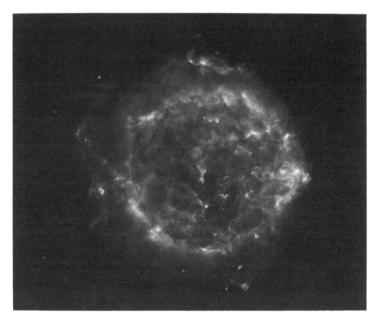

FIGURE 38: A highly detailed radio image of Cassiopeia A. The gauzy interior resembles neither the shell-like appearance of remnants like the supernova of 1006 nor the filled-in appearance of remnants like the Crab Nebula. Sidney van den Bergh has suggested that remnants like Cassiopeia A are the products of Type Ib supernovae. (Courtesy of P. E. Angerhofer, R. Braun, S. F. Gull, R. A. Perley, R. J. Tuffs; NRAO/AUI.)

of the Crab). Runners usually run the second half of a race more slowly than the first, and that might well be the case for supernova remnants. As the expanding shell rammed into surrounding material, it could have swept up some of the slower-moving gas, slowing its own motion in the process. Assuming the supernova did go off in 1680 and had slowed down slightly since, we can estimate the amount of expelled material in the Cas A remnant. Astrophysicist Kenneth Brecher of Boston University calculates a figure of about 10 to 12 solar masses of material, just about what we might expect from a Type II supernova. Estimates of the amount of material needed to produce the X rays seen from the remnant are in good agreement with this figure.

Modern scholarship and modern technology thus suggest that

Flamsteed may have been the last person to witness a supernova in our own galaxy. Like so many things in the history of supernovae, it was a discovery that went unrecognized for centuries.

YOUNG SUPERNOVA REMNANTS

The Crab Nebula and Cas A are two fascinating cases of what happens during the first thousand years after a supernova. By studying other remnants of varying ages, we can piece together a complete picture of the effects of a supernova on its surroundings and compare what we see with what we expect to happen.

What do we expect? Theoreticians describe three distinct stages in the life of a supernova remnant. The first is a stage of free and rapid expansion. Shot outward from the exploding star at speeds of tens of thousands of kilometers per second, an expanding shell of gas, containing at least several times the mass of the sun, balloons into the surrounding space. It encounters little resistance at first and it shines primarily from the energy pumped into it by the explosion and its radioactive by-products. Within a week or two it has inflated to the size of the solar system. We see the beginning of this stage when we watch the first months of a supernova outburst. Supernovae in other galaxies are so distant, however, that we cannot actually see the shell expanding; all we see is the rise and decline of light from the blast. Though we have never been able to observe a supernova for more than a few years after the blast (SN 1987A will no doubt be an exception), the free expansion phase should last for several hundred years at least.

The second stage begins when the expanding shell begins to slow as it meets resistance from the material that surrounds it. The space between the stars is a better vacuum than we can produce in terrestrial labs, but there is still a bit of "interstellar" material there. On the average, a volume of interstellar space the size of a quart milk bottle will contain about 500 atoms of various types (primarily hydrogen), with an occasional microscopic particle of dust thrown in for good measure. If the exploding star had expelled some matter before the blast in the form of a stellar wind, as aging stars are known to do, there may also be a local haze of somewhat denser-than-average gas surrounding the explosion (a "circumstellar" cloud, in technical terms, to distinguish it from the thinner "interstellar" material). We see evidence of that circumstellar gas, for instance, in the slow-moving knots of gas around Cas A.

When the expanding gas from a supernova encounters this surrounding material, it is still moving so rapidly (thousands of kilometers per second) that atoms don't have a chance to move to one side to let it pass. Matter is bunched up ahead of the gas, forming what is technically known as a shock wave. Similar shock waves are produced by nuclear weapons, and this stage of the expansion is often called the Taylor–Sedov phase after two nuclear-weapons theorists, one British and one Russian, who first described it. Heated by the encounter with the material ahead of it, temperatures in the expanding shock rise to millions of degrees, stripping electrons from their atoms. The free electrons, trapped in magnetic field lines in the denser regions of the shock, emit synchrotron radiation at radio wavelengths. The million-degree gas is too hot to emit much light in the visible region of the spectrum, but it naturally emits very-short-wavelength electromagnetic radiation in the form of X rays.

How long a remnant spends in free expansion before the beginning of the Taylor–Sedov phase depends on how much resistance the expanding nebula meets from the material that surrounds the exploding star. And how much resistance it meets depends on how much material there was around the star before it blew up. It also depends on how much material was ejected by the explosion; it is harder to slow down a rolling railroad car than a child's wagon. The time before the Taylor–Sedov phase thus depends on the circumstances of a particular supernova, but estimates range from a few hundred to a few thousand years.

After it reaches this second stage, a remnant pushes its way through the interstellar medium for tens of thousands of years. Gradually the shock slows and cools as it sweeps up more and more interstellar material and radiates away more and more energy. In the third and last stage of its life, bloated to 100 light-years in size, it is beginning to blend with its surroundings. All we see are long wisps and thin sheets of denser-than-average gas drifting slowly away from the long-ago explosion. In a few million years, it will be indistinguishable from the general interstellar material.

Thanks to the diligence of the ancient Chinese and European astronomers, we know the ages of several supernova remnants. If the three-stage theory gives us a reliable timescale, then all the remnants of historical supernovae should be in either the later years of the free-expansion stage, or the early years of the Taylor–Sedov stage. They should consist of shells of million-degree gas moving at speeds of thousands of kilometers per second. Because of their high temperatures, the shells of historical supernovae should not emit much visible light, but they should be intense radio and X-ray sources. That is indeed what we find: most supernova remnants are discovered by the radio waves and X-ray techniques, not by taking photographs of the sky. One of the few exceptions to the rule was the remnant of the supernova of 1604, which Baade, using the position published by Kepler, photographed in 1947.

After the Crab Nebula, the 1604 remnant, and Cas A were discovered, optical and radio astronomers teamed up to locate the remnants of other historical supernovae. In 1952, astronomers Robert Hanbury-Brown and Cyril Hazard located a strong radio source not far from the position given by Tycho in his *De Nova Stella*. Shortly thereafter, Rudolph Minkowski detected an optical image of the remnant of the 1572 star on long-exposure photographs made with the 200-inch telescope: just a few tattered shreds could be seen. The remnant of the brilliant 1006 supernova, similarly, was first detected as a radio source in the early 1970s, and was only photographed by Sidney van den Bergh in 1976.

Even when no visible source is present, radio telescopes can detect the presence of a supernova remnant. Astronomers look for expanding sources of synchrotron radiation that show the round shape we might expect from a blast wave. 3C 58, the remnant of the 1181 supernova, was discovered this way even before the ancient records were known. Sometimes an optical source can later be found: van den Bergh has published photographs of several dozen supernova remnants he and other astronomers have located. But the list of radio remnants outnumbers these by over a hundred. Because supernova remnants are also brilliant X-ray sources, the survey of the sky carried out by the Einstein X-ray satellite between 1978 and 1981 added more candidates to the list.

When we survey the results of these searches, we find that many youthful supernova remnants look just like the spherical shells of debris we expect. Tycho's remnant, Kepler's remnant, and the remnant of SN 1006 are all clearly defined rings in the radio and X-ray pictures, brightest around the edges. They look like rings, rather than filled circles, because our line of sight encounters more emitting material when we look at the edge of a hollow shell than when we look at the center. The sizes of the X-ray- and radio-emitting regions coincide, indicating that both the hot gas and the electrons occupy the same space.

Some of the remnants, however, present more of a puzzle. The Crab Nebula, for instance, appears as a filled-in blob of radiation to radio observers, and as a somewhat smaller blob in X-ray images. 3C 58, although different in shape, is also bright throughout, not just at the edges.

Some astronomers think the two types of supernova remnants arise from different types of explosions. Astronomer Kurt Weiler has named the filled-in type of remnants "plerions" and proposes that the plerions are remnants of Type II supernovae, whereas the hollow-shell type remnants result from Type I supernovae. Indeed the light curves recorded by Tycho in 1572 and Kepler in 1604 both seem to be of Type I, and both their remnants appear to be hollow shells.

But why should the Type II's produce different remains? Type II supernovae leave behind a neutron star; perhaps the neutron star provides the energy needed to fill up and illuminate the interior of the expanding shell. Indeed, in a few of the plerionic remnants, Fred Seward and colleagues at the Center for Astrophysics in Cambridge, Massachusetts, working with X-ray images from the Einstein satellite, have recently found small intense "point" sources of X rays embedded in the surrounding gas. These may be neutron stars feeding energy to the remnant.

On the other hand, some of the variations that we see in supernova remnants may result from differences in the local environment. If a supernova explodes in a region relatively free from surrounding gas, we might expect the remnant to look different than if the explosion occurs in a region adjacent to some dense, resistant interstellar clouds. The distinction between plerions and shell-type remnants is only a first step to understanding, and astronomers will need to do a good deal more work to untangle the effects of heredity and environment on the products of a supernova blast.

OLD SUPERNOVA REMNANTS

Time takes its toll on a supernova remnant. As it forges outward into space, it sweeps up more and more of the surrounding material and, like a runner with a heavy load, it eventually slows its pace. As it slows, it cools, for its high rate of X-ray and radio emission drains its resources of energy. After 20,000 to 30,000 years has elapsed, the shock wave has almost dissipated, and the debris has begun to blend into its surroundings. A number of larger, older remnants give us a picture of this later stage of development.

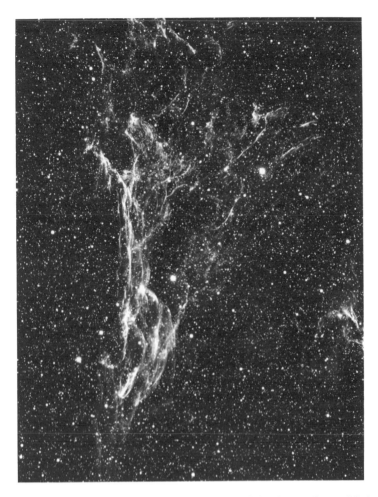

FIGURE 39: The Veil Nebula in Cygnus, a portion of the Cygnus Loop. (National Optical Astronomy Observatories.)

One of the loveliest of these, the Cygnus Loop, is in the constellation of Cygnus, the Swan, high in the evening sky of late summer. Long photographic exposures reveal it as a broken circle of lacy filaments covering a region about five times the diameter of the full moon. This is so large that most telescopes cannot take it all in at one time. As a result, various parts of it are sometimes called by separate names: the Veil Nebula, the Lacework Nebula, the Carrot, and the Network Nebula.

The Cygnus Loop looks enormous from Earth because it is enormous. At an estimated distance of 2500 light-years, the Loop is about 130 light-years from edge to edge. From its speed of expansion we can estimate its age—about 40,000 years, making it one of the older nearby remnants. Another very similar remnant in the constellation of Auriga, Shajn 147, may be a bit older; its filaments are even more delicate and diffuse than those of the Cygnus Loop.

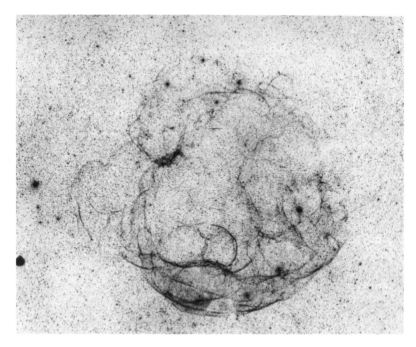

FIGURE 40: Shajn 147, an old supernova remnant, photographed by Rudolph Minkowski. (Palomar Observatory photograph. Courtesy of Sidney van den Bergh.)

There are a number of other well-known supernova remnants that are probably a bit younger than the Cygnus Loop. In the constellation of Gemini, remnant IC 443 appears a bit more energetic and substantial; it may be 20,000 to 30,000 years old. Its more distorted shape probably resulted from a collision with a nearby cloud of very dense gas. Elsewhere, a tattered ring of filaments in the southern constellation of Vela probably formed a mere 10,000 years ago. The Vela remnant may be in the later half of its Taylor–Sedov expansion.

If we compare the old with the young, the ravages of age are apparent. Though the older remnants are easier to spot at visible wavelengths (because cooler, denser gases of the old remnants emit fewer X rays and more light), their radio and X-ray luminosities are

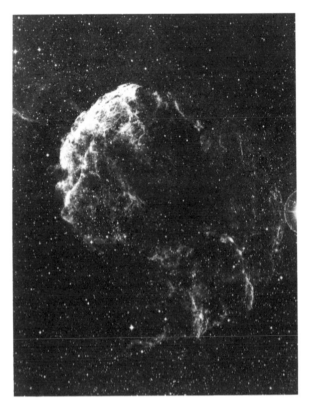

FIGURE 41: IC 443, an old supernova remmant. (Palomar Observatory photograph.)

beginning to decline. The sharply defined boundaries of some of the youngest shell remnants have disappeared, leaving only a loose association of wisps and tatters.

Analysis of the chemical abundances in the remnants shows that the youngest remnants, like Cas A and Kepler's remnant, are rich in the heavy elements we expect to be produced in the interiors of massive stars: oxygen, nitrogen, and sulfur, for instance. The Crab Nebula, always a peculiar one, is unusually rich in helium. In contrast, the oldest remnants like the Cygnus Loop have chemical abundances virtually indistinguishable from normal interstellar gas: nine atoms of hydrogen to one atom of helium, with a smattering of heavier elements. It's not difficult to explain the source of this difference. By the time a remnant ends its Taylor–Sedov stage, it has swept up so much interstellar gas that the enriched debris of the original star is only a small fraction of the whole remnant. The material in the shock has not only been slowed, it has also been diluted.

The exquisite tracery of a remnant like the Cygnus Loop belies the enormous amount of energy that produced it. Consider this: the remnant began, 40,000 years ago, as a wave of material as massive as of ten suns, ejected from an exploding star. In the beginning, it moved at a sizable fraction of the speed of light. Eventually it swept up a hundred suns' worth of surrounding gas. And all that time it has stayed lit with no internal source of energy other than that imparted to it by the original blast. This is pyrotechnics with a vengeance.

In time, however, the fire and fury will be forgotten. In a million years, the Cygnus Loop, its atoms incorporated into the clouds of ambient interstellar gas, will no longer be recognizable. I see it, sometimes, as the turning of a cosmic wheel. The star that gave birth to the supernova remnant was once a cloud of gas itself, and the remnant, scattered and diffused, will someday condense to stars again.

CHAPTER 8

The Eye of the Storm

> The mad things dreamt of in the sky
> Discomfort our philosophy.
> —John Updike, *Skyey Developments*

LOOKING FOR PHANTOMS

Until the late 1960s, the story of supernovae was a murder mystery without a body. Astronomers had seen signs of violence, no doubt about that; but there were no mortal remains. In only one case, the supernova of 1054, was there even a bit of evidence: a strange, faint star in the tangle of the Crab Nebula. Occasional ripples of light emanated from it, stirring up the surrounding gas. To the dismay of observers, its featureless spectrum gave no information about what it was like. Was the Crab star a stellar corpse? If so, what was it composed of? Were there other stars like it in other remnants? There were plenty of conjectures, but no hard facts.

Even before the Crab was identified with an ancient supernova, Baade and Zwicky, in the 1930s, had pointed the finger at neutron stars. Though the notion was quite plausible—one expected stars to shrink dramatically when they died—it generated little enthusiasm among observers. The reason: there seemed no way to test it. A neutron star should be about the same diameter as Manhattan Island, making its surface several billion times smaller than the sun. Only under the most extreme circumstances, with a surface temperature of about a million degrees, could an object that small emit enough energy to be detectable over interstellar distances. No one seriously believed a stellar surface could ever get that hot or, supposing it could, that it would stay that hot for long. So the common wisdom was that once a star collapsed to neutron star proportions it would vanish from our sight. If so, then the mysterious object at the center of the Crab was no neutron star, and there was no prospect of ever seeing one anywhere else.

Other possibilities were even less encouraging. Suppose that the collapse of a massive star went to the ultimate limit, until the star had no diameter at all and its entire bulk was concentrated into a single point at the center. Something this compact—its volume zero, its density infinite—confounds our commonsense notion of matter taking up space and having weight. But such objects were quite possible according to Einstein's general theory of relativity. During World War I, the brilliant German astronomer Karl Schwarzschild, though suffering from a fatal illness contracted at the Russian Front, had worked out the details and sent them to Einstein. Schwarzschild found that the gravitational field of such a collapsed object would be so strong that light emitted near it would be pulled back to the center, and light passing close by would be trapped forever. Theorist John Wheeler, many years later, called Schwarzschild's objects "black holes," an apt name as they neither emit nor reflect any form of radiation. They should be as black as anything we can imagine—and even harder to detect than neutron stars.

To the most skeptical, there was no guarantee that anything remained after a supernova blast. If a star were completely shredded by the blast (as we today believe *does* occur in Type I explosions), then the search for a stellar cinder was as futile as the search for the Holy Grail. In any case, whether a supernova produced a neutron star or a black hole or nothing at all, astronomers despaired of ever performing a postmortem on it. When, quite unexpectedly, they stumbled over a still-warm body, it took some time to realize what they had discovered.

THE DISCOVERY OF PULSARS

Jocelyn Bell was not looking for neutron stars; she was not even looking for stars at all. In 1967 she was trying hard to get her Ph.D. with Anthony Hewish in Cambridge, England. Hewish, an energetic and inventive radio astronomer, along with several colleagues and graduate students, was developing a technique for discovering quasars. Only a few of these powerful sources of radio waves had been recognized at the time, and astronomers knew little about them, other than that they seemed to be billions of light-years away, farther than anything else in the universe. Hewish hoped to distinguish quasars from other sources of radio emission, such as galaxies and supernova remnants, by their distinctive twinkling or scintillation.

It's common knowledge that you can tell a planet from a star because stars often appear to twinkle, whereas planets usually don't. The reason for the twinkling is not in the stars, but in Earth's atmosphere, which is in constant, turbulent motion. Rays of light passing through the atmosphere are deflected to one side or another as they pass through the unsteady air, and as a result the total amount of starlight entering an observer's eye will fluctuate. (You see an exaggerated effect of the unsteady atmosphere when you see the shimmering of the horizon across a desert on a hot day.) The atmosphere does not distinguish light from a star from light from a planet. Thus, both stars and planets twinkle, but planets twinkle less because they are actually small disks of light, rather than starlike points. Each point on the disk of a planet, like an individual star, will appear to twinkle randomly, but when the entire disk is viewed at once, the aggregation of random sparkles blends into a uniform glow. Thus, the total brightness of a planet remains relatively unaffected by our unsteady atmosphere, and it appears as a stable, unwavering light.

Hewish hoped to exploit a similar effect, but on a cosmic scale. Quasars, because they are intrinsically small and exceedingly distant, appear as highly concentrated, starlike sources of radio emission, whereas galaxies and supernova remnants are more spread out, "extended" sources. As radio waves travel from a quasar through interstellar space, they are bent back and forth by passing clouds of interstellar material, just as light rays are buffeted by a turbulent atmosphere. The quasar signal received at Earth should therefore fluctuate randomly, just like a twinkling star. More spread-out sources, like galaxies, should twinkle less or not at all. In 1964, Hewish demonstrated that this was indeed the case, but now he wanted to put the effect to good use by compiling a list of all the suspected quasars he could find. He needed a radio telescope that could be devoted to the search.

It required a special type of instrument. Most radio telescopes of the time were unsuited to the task, primarily because they were not designed to detect the rapid variations in signal strength Hewish expected. A normal radio telescope was simply pointed at a source for several minutes or several hours while it collected all the radio waves coming in. At the end of that time the total amount of energy received from the object could be printed out. This is like opening a camera shutter for a long time in dim light—the long "exposure" makes it possible to detect fainter objects. Conventional radio telescopes, in effect, produced still pictures, not movies. There was no second-by-second record of the strength of an incoming signal. Yet Hewish required just such a record in order to pick out scintillating radio sources. He decided to design and build his own receiving system.

Jocelyn Bell joined Hewish's group in 1965, just about the time construction began. To increase signal strength without requiring long exposures, Hewish had designed an antenna with as large a collecting area as he could. (You use the same principle to make snapshots in dim light: you can open the lens of a camera wider, rather than keeping it open longer.) The Cambridge antenna was so large, in fact, that it had to be built flat on the ground, and it looked quite unlike a conventional radio telescope. It was not a large movable dish, but a set of wires hung from over a thousand slender poles, covering an area of over four acres. In photographs, it looks like a giant colony of clotheslines waiting for wash day. For two years Bell worked with the group, stringing over 120 miles of wire and cable, until, in the summer of 1967, the telescope was ready to survey the sky.

Bell oversaw the day-to-day operation and monitored the accumulating data. As Earth turned, different objects would pass over the antenna, so a continuous record of every signal had to be kept and each source identified. To do this, radio waves picked up by the large antenna were fed to a chart recorder, where a pen traced a record of the signal strength on a moving strip of paper fed from a large roll. Each day, the telescope generated a hundred feet of charts, the output of the receiver showing up as a line running from one end of the paper to the other. And each day, Jocelyn Bell personally examined the records, looking for quasars.

By the time she had scanned a few hundred feet of paper, Bell became familiar with the different appearances of the sources. A particularly intense source would produce a large jog in the line, and a twinkling, or scintillating source would show up as a closely spaced series of wiggles. She learned to reject signals due to man-made interference: artificial satellites, airplane radar, automobile ignition, even a malfunctioning home appliance. But there were a few odd cases. "Six or eight weeks after starting the survey," she later recalled, "I became aware that on occasions there was a bit of 'scruff' on the records, which did not look exactly like a scintillating source, and yet did not look exactly like man-made interference either. Furthermore, I realized that this scruff had been seen before on the same part of the records, from the same part of the sky." Here was an unforeseen puzzle.

The "scruff" was a rapidly fluctuating signal of some sort, the wiggles of the chart pen all squeezed together by the rapidity of the fluctuation. On November 28, 1967, Bell recorded the signal on a strip of very rapidly moving paper to spread out the wiggles and inspect them. "As the chart flowed under the pen, I could see that the signal was a series of pulses, and my suspicion that they were equally spaced was confirmed as soon as I got the chart off the recorder. They were one and one-third seconds apart."

Such uniform and rapid blips were unheard of from any natural object. It was natural to attribute them at first to some artificial source. But Bell and Hewish could connect no earthly interference with the signals, try as they might. There were no radars in all of Britain that produced pulses one and one-third seconds apart. All the evidence pointed to a celestial origin. Terrestrial signals, for one thing, usually keep time with the sun: you'd expect a radio station or an electric razor to be turned on at about the same time every day. In contrast, the "bit of scruff" appeared on the chart paper about four minutes earlier every day. This is just the same behavior as the stars: the constellations rise earlier by four minutes a day as a result of Earth's yearly motion around the sun.

Whatever was causing the pulses, then, it was something beyond Earth, and very likely beyond the solar system. Yet why were the pulses so unnaturally regular? Precise measurements revealed that the pulses repeated every 1.33730113 seconds with a high degree of accuracy. This rivaled the performance of the best laboratory clocks. "We had to face the possibility," said Hewish, "that the signals were, indeed, generated on a planet circling some distant star, and that they were artificial." Could someone out there be trying to tell us something? Or was this some sort of navigational beacon for interstellar spacecraft?

It was late December 1967 and the Cambridge group still had not published their discovery. Unless they could attribute it to some natural cause, this eccentric radio source could make fools of them all. No one wanted to go to press and later find that the Cambridge radio telescope had spent the fall examining a sparking toaster or a neon sign. And if it *was* a message from some alien civilization, who would believe them? "Here I was," lamented Bell, "trying to get a Ph.D. out of a new technique and some silly lot of little green men had to choose my aerial and my frequency to communicate with us."

But just before Christmas she saw some more scruff on the chart records, at a different time of day and from a different place in the sky. It turned out to be a second pulsating radio source; within a few weeks two more had been confirmed, all pulsating at different rates. With that, the little green men hypothesis went into the dustbin of history: surely four different civilizations, light-years apart, could not send out the same sort of tedious signal at Bell's frequency. It was time to publish. Regularly pulsating radio sources, pulsars for short, were demonstrably natural in origin. No one understood just what they were, but when Hewish, Bell, and their collaborators published news of the discovery in the February 24, 1968 issue of *Nature*, they raised the possibility that the pulses came from some sort of dense, small object like a white dwarf or a neutron star. Only a small object, they believed, could alter its light output so drastically over so short a time.

Bell received her degree and went on to teaching and research, none as epoch-making as her serendipitous discovery of "scruff." For his work on radio astronomy, as well as for the codiscovery of pulsars, Hewish shared the Nobel Prize in 1974 (with radio astronomer Martin Ryle).

THE NATURE OF THE BEAST

I was just a graduate student at the time, but I can well recall the almost universal excitement generated by the announcement of the first pulsars. Radio astronomers rushed to observe the new objects, confirming the stellar distances of the sources and the uncanny regularity of the pulses. Within weeks, as the implications of the discovery became clear, theoretical interpretations of the radio emission began to appear in the professional journals. For some time, confusion reigned, but within a few months it seemed certain that only one kind of explanation would do: the pulsars were rapidly rotating neutron stars.

Consider the possibilities. Astronomers recognize three ways to produce regular variations in the light from a star. The first occurs among certain binary stars called eclipsing binary systems. If the orbit of the pair is aligned with our line of sight, one star will pass in front of the other each time it goes around. We thus see a brief eclipse as one star blocks off the light from the other. The closer the two stars are to one another, the more rapidly they orbit. In an extreme case, when two neutron stars orbit so close that their surfaces graze one another like square dancers in a do-si-do, eclipses can occur more frequently than once a second. But such a situation seemed highly contrived; moreover there were indications that a pair of stars so closely separated would tear each other apart. Pulsars simply blinked too rapidly to be eclipsing binaries.

A second type of object that produces variations is a pulsating (or vibrating) star, a star that swells and shrinks like a beating heart. Cepheid variables brighten and dim in such a manner, as do many other types of stars. The rate at which they pulsate is determined by their size and by the density of the material that makes them up, just as the pitch of a ringing bell depends on how big it is and what type of metal it is made of. Cepheids, for instance, which are diffuse, giant stars, bloat and shrink over timescales of days to months. White dwarfs, far smaller and denser, can vibrate as fast as about once a second. And neutron stars, even smaller and denser than white dwarfs, naturally oscillate thousands of times a second.

In early 1968, it seemed possible that Bell's first pulsar, now called CP1919 (the letters stand for "Cambridge Pulsar," the numbers for its position in the sky), might be produced by a vibrating white dwarf. Its period of vibration was a little over a second, just as one might expect. But the subsequent discovery of pulsars that pulsed more rapidly than a second made the vibrating white dwarf hypothesis untenable. If pulsars, as it turned out, vibrated too rapidly to be vibrating white dwarfs and too slowly to be vibrating neutron stars, then stellar vibrations were out of the running. No other type of stars had vibration periods in the observed range.

The third possibility was that a rapidly spinning star was sending out a narrow beam of radio waves into space like a gigantic light-house. Every time the beam crossed our line of sight we would see a brief flash, the rest of the time, virtually nothing. The idea was appealing in its simplicity. If this was the cause of pulsars, the spinning object could be nothing but a neutron star.

Only a neutron star could spin fast enough. The sun spins at the leisurely pace of once every 26 days. If it were to rotate faster, it would start shedding matter like mud from a spinning tire, and would fly apart completely if it were rotating once every 3 hours. Smaller, denser stars than the sun exert a stronger force of gravity on objects at the surface, and therefore can spin faster without coming apart. A white dwarf remains intact up to several revolutions a second. But a neutron star has such a strong surface gravity that it can spin a thousand times faster than a white dwarf. There was little doubt that spinning neutron stars could comfortably account for the pulse rates of all the known pulsars.

Less than six months after the first paper on pulsars, Thomas Gold of Cornell University published an article in *Nature* in which he argued for the spinning neutron star explanation. Everything seemed to fit together neatly. Gold's ideas meshed not just with the observations of pulsars, but also with the general picture of the origin of neutron stars. We would expect neutron stars to be spinning rapidly, he noted, because they come from the collapse of ordinary stars. Like a skater drawing her arms closer to her body, a star should spin faster and faster as it gets smaller. Were the sun to shrink to neutron-star size, for instance, it would be rotating once every few seconds, just about right for a pulsar.

If pulsars were neutron stars, we could understand not only the rapidity of their pulses, but also their impressive regularity. A neutron star would behave like a massive flywheel mounted on frictionless bearings. If you have ever tried to stop a rapidly spinning bicycle wheel with your hands, you have some idea of how difficult it is to slow down a rotating object. Imagine the wheel were more massive: a trillion times denser than solid steel . . . ten thousand times larger. How could you stop it then? The answer is: only by applying an enormous force. But a spinning star encounters practically no resistance from its surroundings. That is why the beam of light from a pulsar sweeps by us on such a precisely rigid schedule. The spinning neutron star acts as a well-regulated mainspring for the pulsar clock.

In his paper Gold also suggested how a spinning neutron star could generate a beam of radio waves. The necessary ingredients were (1) a strong magnetic field and (2) rapidly moving, electrically charged particles. Collapsed stars should have both. The magnetic field strength at the surface of a star should increase as the star shrinks, because lines of magnetic force present at the start are bunched together over a smaller area. If the sun were to shrink to the dimensions of a neutron star, its surface magnetic field strength would increase about ten billion times. It is also likely that such a dense star would be surrounded by a hot gas, with electrons stripped from atoms and flying around at high speeds.

The rapidly rotating, intense magnetic field of a neutron star will snag electrons from the surrounding gas, accelerating them to speeds approaching that of light. The trapped moving electrons whirl around the magnetic field lines, which are themselves rotating with the spinning star. As in synchrotrons, the whirling electrons give off electromagnetic radiation. The result is a narrow beam of intense radio waves that whips around like a beacon, crossing our line of sight on each revolution of the neutron star.

One word of caution: I have presented here a highly simplified description of a complex process. The radiation mechanism for pulsars is still not well understood, and there are several schools of thought on the details. How is the magnetic field oriented with respect to the axis of rotation of the spinning neutron star? Does the radiation come from the poles of the star or from a region in the plane of the equator where the magnetic field moves with the speed of light? Is the radiation emitted in a narrow cone or a wedge encircling the equator? Details of the process are difficult to analyze and difficult to test against observations.

But the fundamental elements of Gold's scenario for pulsars have stood the test of time. A rapidly spinning neutron star with a strong magnetic field is present in all the successful models. The explanation of the mystery, coming so soon after the discovery of pulsars, was a remarkable feat. More remarkably, perhaps, Gold's conclusions had been anticipated even *before* the discovery of pulsars.

In 1964, Russian physicist V. L. Ginzburg had noted that a collapsed star should have a strong magnetic field. His compatriot N. S. Kardashev, shortly thereafter, argued that a rapidly rotating magnetized neutron star might explain the high magnetic fields in supernova remnants. And just a few months before Jocelyn Bell's discovery of CP1919, Italian Franco Pacini published a paper in *Nature* that discussed rotating magnetic neutron stars in the context of nebulae like the Crab. For years astronomers had been looking for a source of energy to power its synchrotron radiation. What kept it shining? What energized the activity we saw near its center? Could it be a rotating neutron star? Pacini thought so. But neither he, nor Kardashev, nor Ginzburg had proposed looking for radio pulses from the Crab Nebula.

The Crab, as it happened, provided the missing link between Pacini's speculation about the energy source of supernova remnants and Gold's conjecture about the nature of pulsars. In November 1968, David H. Staelin and Edward C. Reifenstein, using a 300-foot dish antenna at the National Radio Astronomy Observatory in Green Bank, West Virginia, discovered a pulsar, designated NP 0532, right in the heart of the Crab. It flashed more rapidly than any yet discovered, almost 30 times per second. This was too fast to be accounted for by a vibrating white dwarf; it could only be a rotating neutron star.

Which star was it? The radio positions were not precise enough to decide whether the pulses were coming from the mysterious star at the center of the Crab or some other object in the vicinity. If the central star was the Crab pulsar, then it, too, should be flashing at 30 times a second. No previous photographs would have revealed the pulses, however; astronomers (like radio astronomers) normally take long-exposure still pictures, not movies, of stars. With modern electronics, however, it was not difficult to get a running record of the light from the central star. In January 1969, John Cocke, Donald Taylor, and Michael Disney, using a small telescope at Kitt Peak National Observatory in Arizona, recorded pulses of light, perfectly synchronized with the radio pulses, coming from the central star of the Crab remnant. Within days, other groups of observers confirmed the result. The central star of the Crab Nebula, already a suspected neutron star, was evidently a pulsar.

The conclusive evidence was yet to come. According to Gold's mechanism for pulsars, the spin of a magnetized neutron star should gradually slow down. The reason was that a neutron star supports itself without an independent energy source. Because its nuclear reserves are spent, a neutron star must tap into its energy of rotation to produce light and radio waves. In this respect, a pulsar is like a large turbine generator in an electric plant. If you turn off the steam, the generator will continue to spin for quite some time. But as households draw electricity from the generator, it slows in response, the lights dim, and eventually all power production ceases.

At the time of formation, right after a supernova blast, a pulsar

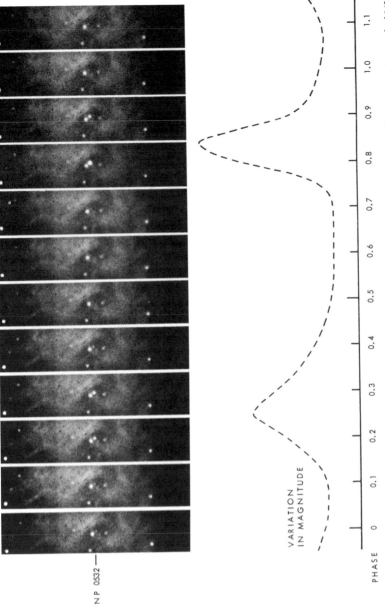

FIGURE 42: Optical variations of the Crab Nebula pulsar, NP 0532. The entire sequence of photographs spans an interval of 1/30 of a second. (National Optical Astronomy Observatories.)

spins most rapidly—there is considerable debate about the precise figure, but estimates run to several hundred times a second. It slows down as it radiates energy, with the slowdown progressing most dramatically at first. In a thousand years a pulsar slows to 30 times a second, but after ten thousand years it still rotates about 10 times a second or so. All the while it steadily fades until, several million years after its birth, it goes dark. Only a cold neutron star remains, virtually undetectable by earthbound astronomers.

In October 1968, just before the Crab pulsar was discovered, astronomers discovered a rapid pulsar in the Vela supernova remnant, a network of wispy nebulosity in the southern sky. It was less luminous than the Crab pulsar and it pulsed more slowly: about 12 times a second. According to ancient records, the Crab was over nine hundred years old; the Vela remnant was considerably older. As predicted, the younger pulsar was more rapid and more energetic. Someday, as it spun out more and more of its energy, the Crab pulsar would look like its cousin in Vela. The pieces of the puzzle were beginning to fit together.

Two months later, the case was virtually closed. In Arecibo, Puerto Rico, astronomers had recently completed the world's largest radio telescope. The interior of a natural valley was lined with wire mesh, making a dish a thousand feet in diameter. Suspended above the dish was an antenna assembly the size of a football field. It was a massive structure whose size was all out of proportion to the feeble signals it was designed to detect. The huge collecting area, however, made it possible to study the pulses from the Crab in exquisite detail. In December 1967, the Arecibo telescope obtained a lengthy sample of signals from a newly discovered pulsar.

There was a clear trend in the data. The interval between pulses from the Crab was increasing slightly day to day. It was a minuscule effect, just a few millionths of a second each year. That, however, was just as predicted. The pulsar at the center was feeling its age; it was slowing down. When astronomers calculated just how much energy the Crab pulsar was losing each year, it turned out to be a familiar figure: it was the energy required to keep the Crab Nebula shining (see Chapter 7).

There was no room for doubt any more. Baade and Zwicky's crazy suggestion of 1934 didn't seem crazy any more. And the Crab Nebula didn't seem as puzzling. Pulsars, rotating neutron stars forged in the heat of supernovae, could explain it all.

PULSARS, NEUTRON STARS, AND SUPERNOVA REMNANTS

Since Jocelyn Bell's discovery of "scruff," hundreds of pulsars have been discovered by radio astronomers. Most are scattered around the arms of the Milky Way, within a few thousand light-years of the sun; others have been discovered in the Small and Large Magellanic Clouds, two smaller galaxies about 160,000 light-years away. Progress toward an understanding of the relation between pulsars, neutron stars, and supernovae continues, but at a much less frenetic pace than during 1967 and 1968.

One clear fact that has emerged from all the research is that not all neutron stars are pulsars. A young, rapidly spinning neutron star may produce optical and radio pulses if it has a strong magnetic field, but we know of no reason why this must always be the case. Moreover, even if a neutron star begins as a pulsar, it does not keep producing pulses forever. As a pulsar ages, its optical radiation disappears first, probably within a few thousand years. The radio pulses weaken with age, too, and no pulses should be detectable at all after several million years. As our galaxy is more than a thousand times older than that, most of the neutron stars that have formed over that time remain dark and forever unseen by our telescopes.

In our galaxy, for instance, only two pulsars have been detected at optical wavelengths: the Crab pulsar, and the Vela pulsar. They remain, significantly, the youngest pulsars in our galaxy, and they appear to be spinning very rapidly. We may conclude that a pulsar older than Vela does not produce enough light to be noticeable. There's the same general pattern in the radio emissions: the more rapid pulsars, presumably younger, are more luminous. The slower pulsars, presumably older, are fainter, suggesting that there are even slower rotators that we just do not notice.

Furthermore, even a young, rapidly spinning neutron star may not be a noticeable pulsar. Unless the beam of radiation it produces happens to sweep across our line of sight, we will see no pulses from it. If the spin axes of neutron stars are oriented randomly, our chances of seeing a pulsar depend on how broad its beam of radiation is. Because the precise mechanism of pulsar radiation is still a matter of debate, it's hard to say exactly what fraction of young neutron stars produce beams that hit Earth, but estimates are that the number is about 30%. Thus, only one in every three spinning radio sources will appear as a pulsar to us.

If pulsars are spinning neutron stars, and if neutron stars form in supernovae, we might expect to find many pulsars located near supernova remnants. Therefore, it is puzzling that less than a dozen of the hundreds of known pulsars are in or near supernova remnants. Among the remnants of historical supernovae, only one, the Crab Nebula, contains a recognized pulsar. Tycho's and Kepler's remnant, along with the remnant of the 1006 supernova, have no detectable central star of any sort.

At first sight, these facts challenge the notion that supernovae produce neutron stars. But the absence of a supernova remnant near a pulsar is not altogether unexpected. Even if the pulsar formed in a supernova explosion, the remnant should only be visible for a few tens of thousands of years. The pulsar, on the other hand, may be detectable for millions of years. The majority of the pulsars we see may simply have outlived the supernova remnants that surround them.

The absence of a pulsar in at least some supernova remnants is also easy to understand. Only about half the supernovae, the Type II events, produce neutron stars. From the observations made by ancient astronomers, we know that Tycho's and Kepler's star were Type I supernovae. Both these remnants look like hollow shells at radio and X-ray wavelengths (see Chapter 7), a characteristic we generally identify with the remnants of Type I blasts. There is, as we might expect, no trace of a neutron star at the center of these remnants. From the appearance of the remnant of the supernova of 1006, that was also a Type I blast. Again there is no pulsar in the remnant.

The Crab Nebula, in contrast, appears as a plerion, or filled shell. It does contain a pulsar. We are not certain that there is a connection between the pulsar and the appearance of the remnant, but it seems likely. Perhaps a pulsar in a supernova remnant heats gas around it and makes it emit X rays and radio waves. Or perhaps a pulsar in a supernova remnant actually fills its surroundings with gas spun outward from its surface. Whatever the cause, pulsars seem to be associated with filled-shell remnants if they are associated with any remnant at all. Perhaps Type II supernovae produce filled shells and neutron stars, and Type I supernovae, hollow shells with no star at the center.

X-ray studies of supernova remnants support the idea that only

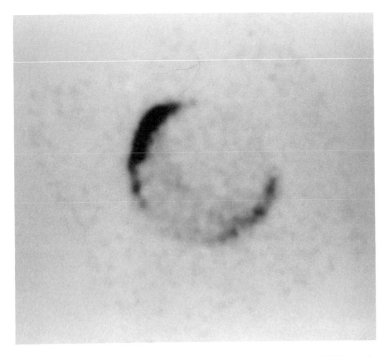

FIGURE 43: A negative X-ray image of the remnant of the supernova of 1006 produced by the Einstein X-ray satellite. The emission comes from a hollow shell of hot gas, and there is no pulsar in the remnant. (Courtesy of Fred Seward, Harvard–Smithsonian Center for Astrophysics.)

the filled-shell remnants contain a neutron star. Pulses from the Crab pulsar were recorded by rocketborne X-ray detectors in 1969. The Einstein Observatory, a large orbiting X-ray telescope that operated between winter 1978 and spring 1981, followed up these early detections by producing images of over 40 supernova remnants in the Milky Way Galaxy and another 25 in the Small and Large Magellanic Clouds. On images of the Crab Nebula taken with the Einstein detectors, the pulsar shows up as a concentrated point source embedded in the indistinct body of the nebula. Other pulsars, researchers speculated, could be recognized as "hot spots" embedded in similar filled-in remnants.

The first neutron star of this sort was detected on Einstein images by Philip Gregory and Gregory Fahlman of the University of British

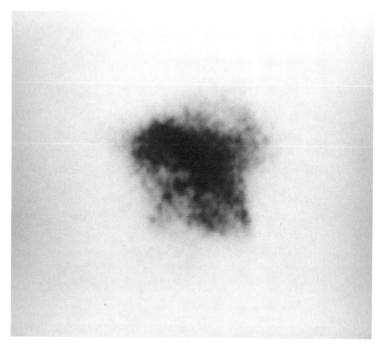

FIGURE 44: A negative X-ray image of the Crab Nebula, produced by the Einstein X-ray satellite. This is a plerion, or filled-shell remnant. A dark spot at the lower left of the remnant marks emission from the Crab pulsar. (Courtesy of Fred Seward, Harvard-Smithsonian Center for Astrophysics.)

Columbia. At the position of a previously known filled-shell radio source named CTB 109, they saw a well-defined round patch with a sharp starlike object in the center. CTB 109 had not been recognized as a pulsar before, but, on reexamination, it clearly was. Its pulse rate, radio observers quickly discovered, was almost 7 seconds, but the interval between pulses was not constant, nor could they see it lengthen like the Crab. Instead it decreased for about 20 minutes, then increased for another 20, repeating the cycle over and over.

This remarkable behavior had a relatively straightforward explanation: CTB 109 was a binary star, one member of which was a pulsar. As the pulsar orbited its companion, it alternately moved away from us and approached, causing pulse arrival times to increase and decrease. The intense spot of X rays in the Einstein image may not have come directly from the pulsar itself, but from hot material flowing from the companion star onto the surface of the pulsar.

CTB 109 is a bit of a curiosity. Only a handful of binary pulsars are known. One might think that an explosion as severe as a supernova would totally disrupt a pair of stars. It would certainly incinerate Earth, were the sun to do the same. Most pulsars, in fact, are observed to be traveling through the galaxy at higher velocities than the average star. The high velocity of such a pulsar can be explained if it was originally a member of a binary system that was torn apart by a supernova. A pulsar, ejected from an orbiting pair like a stone from a slingshot, comes out at high velocity. Yet, if CTB 109 is representative, some binary stars (though probably not a large fraction) do remain together after the blast.

One remarkable pulsar recognized on Einstein images was discovered in February 1984 by David Helfand of Columbia University, and Rick Harnden and Fred Seward of the Harvard–Smithsonian Center for Astrophysics. Officially known by its position in the sky, 0540–69.3, it has earned a reputation as the "Crab Twin," or the "50-millisecond pulsar." The pulsar is located in the Large Magellanic Cloud, not far from the site where supernova 1987A was to flare up in February 1987.

The pulse rate of 0540–69.3—20 times a second—is between that of the older Vela pulsar and the younger Crab. Astronomers John Middleditch and Carl Pennypacker have discovered optical pulsations from 0540–69.3, making it only the third optical pulsar known. Its light rises and falls in intensity in two swift bursts, just like the Crab pulsar. And the surrounding remnant seems Crab-like as well. Though it is ten times more distant than remnants in the Milky Way, astronomers can detect a circular nebula surrounding it on both X-ray and optical photographs. The remnant seems to consist of a bright ring, a bit smaller than the Crab Nebula, surrounded by a larger cloud of gas rich in oxygen. "This system," writes Fred Seward, "is so much like the Crab that we can finally say the latter is no longer unique."

Among the other supernova remnants surveyed by Einstein are a number that show bright spots at the center with no corresponding pulsar. One of these is 3C 58, the remnant of the supernova of 1181. These bright spots may be pulsars that are not beaming their signals in our direction. To make matters more puzzling, however, there are also several filled-shell remnants, which show no central source at either X-ray or radio wavelengths. Perhaps their pulsars are less ac-

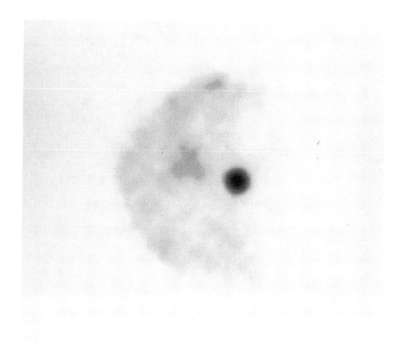

FIGURE 45: A negative X-ray image of CTB 109, produced by the Einstein X-ray satellite. The dark spot near the center of the remnant marks the central neutron star, a pulsar that is a member of the binary system. The X rays probably do not come directly from the pulsar, but from hot gas flowing from one star to the other. (Courtesy of Fred Seward, Harvard–Smithsonian Center for Astrophysics.)

tive than normal, producing radiation that is too faint to detect, even though it energizes the interior of the remnant. Or perhaps some other process is responsible for these unusual objects. We simply do not understand enough about how pulsars produce X rays and radio pulses to know for sure. New X-ray satellites to be launched in the 1990s will no doubt provide additional data to work with.

FROM NEUTRON STARS TO BLACK HOLES

Surely neutron stars must rank among the most bizarre of nature's creations. To make a neutron star, an amount of matter comparable to that in the sun—which is almost a million miles in diameter—must be crammed into a volume only slightly larger than a terrestrial mountain. It is incomprehensibly dense. A speck of neutron star stuff as heavy as a luxury liner could fit comfortably inside a printed letter "O" on this page. Matter this compact loses its atomic identity. Electrons blend with protons; intervening space effectively disappears. A neutron star has been likened to a single giant atomic nucleus, a ball of 10^{57} neutrons, packed like marbles in a jar, with no room left for adjustment. (The analogy isn't totally apt, for nuclear particles don't really behave like rigid spheres, but it will do here.)

Likening neutron stars to giant atoms, however, gives only an inkling of how odd they really are. Neutron star stuff, according to calculations, should behave like a superconductor: transporting electrical current and heat with no resistance at all. Superconducting materials are laboratory curiosities to most scientists. Until recently they could only be produced at temperatures near absolute zero, the coldest temperature there is. Within the past year, physicists have developed exotic alloys that retain superconductivity at temperatures approaching, and perhaps exceeding, room temperature. It is remarkable to recognize entire stars achieving properties that require such diligence and care on a human scale.

A neutron star is anything but delicate. Something as small and massive as a neutron star has an intense gravitational field at its surface. If one could survive the crush, a 200-pound human would tip the scales at over 10 billion tons. A golf ball dropped on the surface would strike with the impact of an atom bomb. Add to this hostile picture a magnetic field a trillion times stronger than that which orients compass needles on Earth and you have an impression of nature pushed to the extreme.

Yet it is possible to imagine objects even more dense than neutron stars. This state of affairs, first described in detail by Karl Schwarzschild, should result if a star like the sun were compressed just a bit more than a neutron star. Pack the sun into a ball a mere 6 kilometers in diameter (a neutron star is about 3 times larger), and, according to Schwarzschild's work, it will shrink with no further hesitation to a single point in space. The critical size, which is larger for more massive stars, is called the Schwarzschild radius: a star compressed smaller than its Schwarzschild radius cannot avoid collapse, no matter what it is made of. The outlandish product of such collapse, a point of infinite density containing the mass of an entire star, is called a black hole.

Mathematically speaking, this point of infinite density is regarded as a "singularity"; physical theory breaks down trying to describe it. In practice, however, the singularity at the center is hidden from view, and we regard the black hole as including everything within the Schwarzschild radius. We cannot see anything, in fact, that lies inside the Schwarzschild radius of a black hole. The reason is that the gravity of a black hole is so great. It is not too difficult to understand why: Imagine that you are on Earth and throw a ball upward, it will soon come back to Earth, pulled back by the force of gravity. If you throw the ball harder, it will return after a longer time. And if you throw it hard enough, faster than a speed called its "escape velocity" (about 11 kilometers per second on Earth), it will coast away from the planet altogether. Now imagine increasing the mass of Earth until it is as great as that of the sun; the force of gravity will increase a millionfold, and you will have to throw the ball faster to get it to leave the planet. Then imagine squeezing the planet smaller and smaller. Surface gravity will increase further, driving the escape velocity still higher in the process. A body small enough and massive enough will have an escape velocity equal to or greater than the speed of light, and nothing will be able to escape it. A ball, which simply can't be made to move faster than light, will return quickly to the surface of such a body no matter how hard it is thrown. Even light emitted upward will bend and come back down.

A black hole, then, is a body so compressed that its escape velocity exceeds the speed of light. For such a body, its Schwarzschild radius is a one-way barrier that separates the black hole from the external universe. Light and matter can pass into it, just as meteorites from the heavens can fall to Earth. But once inside, nothing can get out. A black hole cannot, in principle, emit light or any other form of radiation from inside the barrier. And thus we can't look inside it, no matter how hard we try or how clever we are.

Though such behavior sounds more like science fiction than fact, we have reasons to believe that black holes really do exist, and that some, at least, are formed in supernovae. Their progenitors are stars that have masses of greater than 20 or 30 times the mass of the sun. They are exceedingly rare, but there are a number of such objects known in our galaxy and in other galaxies. In their later stages of evolution, these stellar heavyweights may form iron-rich cores that are simply too massive to become stable neutron stars. Just as Chandrasekhar had shown there was a maximum mass for white dwarfs (about 1.4 times the mass of the sun), beyond which they simply cannot support their own enormous weight, so later workers have shown that there is a maximum size for neutron stars, in roughly the same range. There is an uncertainty in the exact figure, because the behavior of matter at nuclear densities is not fully understood, but there is no question that neutron stars more massive than several times the mass of the sun simply cannot resist the pull of their own gravity.

If this is so, it is possible that collapsing stars might produce cores too massive to support themselves. Or alternatively the weight of such a supermassive star, crashing down during the initial frantic moments of collapse, might push an otherwise stable neutron core inside its Schwarzschild radius. In either case, the ultimate collapse of the core of the star would be irreversible. A black hole would be formed. The problem, it seems, is not to produce black holes, but to detect them.

Nothing is blacker than a black hole. That being the case, how could one ever hope to verify that black holes are indeed formed in supernova blasts? In fact, how could one verify that black holes exist at all? The prospects of seeing a single black hole seem utterly discouraging, as the only effect it has on the universe around it is its gravitational attraction. If black holes were formed in splendid isolation there would, indeed, be virtually no way to tell they were there. But a black hole can sometimes be detected by its effect on a nearby companion. Normal stars are, as often as not, found in pairs. Some survive the supernova blast: we know of neutron stars, like the pair in CTB 109, that are binary. Might there also be some black holes that have normal stars orbiting around them?

Picture a binary system, one member of which is a black hole and the other a normal star. In time the normal star will expand to giant or supergiant size and shed matter onto its black hole companion. (It's not an unlikely scenario. Such mass transfer is seen in ordinary binaries; it's a popular mechanism for Type I supernovae; and it has been observed in neutron-star systems as well.) If the material from the normal star simply disappeared into the black hole, we'd never know it. But it doesn't. Just as water leaving a bathtub doesn't go straight into the drain, infalling material from a giant star does not plummet directly toward the black hole. Instead it spirals inward, forming a whirling mass of hot gas called an accretion disk.

Material does not whirl down into an accretion disk as gently as water flowing out of a tub, however. Pulled inward by the immense gravitational force of a black hole, infalling gas slams into the material already in the disk with all the energy of a nuclear explosion. The scene is one of unimaginable turmoil, with temperatures well above the million-degree mark. At these temperatures, gases emit copious amounts of X rays. These X rays, generated in the accretion disk, provide a signal of the presence of a black hole at the center. Thus, though we never see the black hole itself, we can detect its influence on material in its vicinity.

One prime candidate, named Cygnus X-1, seems to be associated with a blue supergiant star known to astronomers as HD 226868. No one suspected that the star was particularly out of the ordinary until 1971, when an X-ray satellite named Uhuru detected a strong source of radiation in the constellation of Cygnus. The Cygnus X-1 signal came from the vicinity of HD 226868, but the unique nature of the object was not realized for certain until astronomers had a detailed look at the spectrum of the star. Doppler shifts revealed that HD 226868 was constantly on the move, regularly advancing toward us and then receding, as it orbited an unseen companion every 5.6 days.

From the variations in velocity, the mass of the companion star could be estimated: about 10 times the mass of the sun. Any normal star 10 times the mass of the sun would shine brightly enough to be seen. But the companion gives no light. It could be a neutron star or white dwarf too faint to be seen, but no neutron star or white dwarf could be this massive. It must, we conclude, be a black hole. We know it is there, but we see it only indirectly, as X rays, produced by matter streaming from the distended body of HD 226868 into the accretion disk around the black hole. No supernova remnant is seen at the site of Cygnus X-1, and it seems likely that, if the black hole did indeed form in a supernova, the event occurred long before any records were kept here on Earth.

An even stranger object, possibly a black hole, takes its name, SS 433, from the initials of astronomers Bruce Stephenson and Nicholas Sanduleak (SS), who included it in a 1977 listing of stars with unusual spectra. As with Cygnus X-1, however, it was independently studied at other wavelengths before its significance was fully understood. At the time of Stephenson and Sanduleak's survey, radio astronomers were already looking at the region around SS 433. The object of interest was W50, an odd supernova remnant that seemed to have a strong pointlike source of radio waves at the center. X-ray observers, using the British Ariel V satellite, later recognized a small variable source of X radiation at the same position.

Astronomers rarely devote a lot of telescope time to scrutinizing an object unless they believe they can learn something important from it. In this case, optical, X-ray, and radio astronomers with an interest in supernovae began to compare notes. Their different lines of research seemed to point in the same direction, and they turned their attention to the inconspicuous star in the SS catalog. When they observed the spectrum in detail and on a regular basis, they found it quite remarkable. Most conspicuous was a pattern of bright emission lines from hydrogen, which indicated that the star was surrounded by some sort of hot, thin gas. The presence of such a cloud was not especially unusual. But the lines were triple. There was one set of very strong lines at the normal wavelengths of hydrogen, along with a pair of weaker lines flanking each hydrogen line, one far to the red, one far to the blue. It was possible to imagine that each pair of lines arose from two clouds of gas, perhaps opposite sides of a shell, moving with high speeds. Yet if the shifts in the lines were Doppler shifts caused by the velocity of the gas, the cloud was moving faster than the shell of a supernova—over 40,000 kilometers per second. And then where did the third, unshifted set of lines come from? Nothing like it had ever been seen.

More remarkably, the Doppler shifts of the lines in each pair seemed to change with time. Astronomer Bruce Margon, who conducted the pivotal studies of SS 433 in 1978, discovered that over a 164-day period the blueshifted lines would shift to the red and back to the blue again, whereas the redshifted lines did just the opposite. The wavelength of the third emission line between the pair did not seem to change much at all. The star seemed to be coming, going, and standing still all at the same time. How could one explain this evident contradiction?

Most astronomers are convinced that the red- and blueshifted lines come, not from an expanding shell, but from two high-velocity jets of gas squirting away from the star in opposite directions, like water from the nozzle of a high-pressure lawn sprinkler. The axis of the jets doesn't stay fixed in space, but "precesses," or wobbles, just as the axis of a top wobbles as it spins on the floor. The period for this precession is about 164 days. When a jet is pointing in the general direction of Earth, we see the gas coming toward us at the highest velocity. As the orientation of the jet changes, the observed velocity changes until the jet is pointing generally away from us, after which it begins to move toward us again. Because each jet produces its own set of lines, the contrary behavior of the two moving sets of spectral lines arises naturally from the two oppositely moving streams of material: when one jet is pointing toward us, the other is pointing away.

How are the two jets produced? Here the agreement is less universal. We know that SS 433 is actually a binary star. There is a small varying Doppler shift of the third set of strong, almost stationary, spectral lines, which repeats every 13 days. If SS 433 consists of a black hole with a companion orbiting around it, the companion may shed matter into the black hole, forming a large accretion disk around

FIGURE 46: An X-ray image of SS 433. The central dot may be X radiation from an accretion disk around a black hole. The two blobs of intense emission on both sides of the central object come from gas struck by two jets emanating from the object. The regions look triangular because the jets precess (wobble), sweeping out a cone-shaped region in the surrounding space. (Courtesy of Fred Seward, Harvard–Smithsonian Center for Astrophysics.)

it. The ejected matter falls at such a high rate that the black hole cannot swallow it rapidly enough, and the resulting surplus matter may be ejected in a pair of narrow jets in a direction at right angles to the disk. The accretion disk orbits around its companion every 13 days, producing the strong third set of hydrogen lines, while the streaming gas in the jets produces the lines that change every 164 days as the accretion disk varies its orientation to our line of sight.

Astronomers are not yet in agreement as to whether the star at

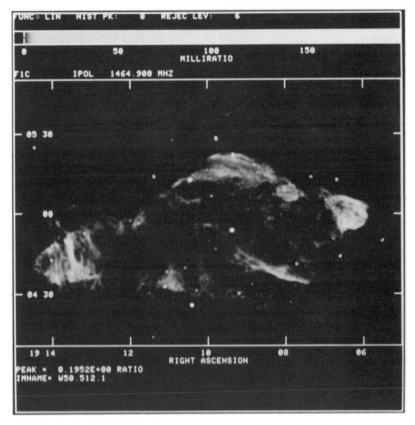

FIGURE 47: A radio image of the supernova remnant W50, SS 433 is in the center of the remnant. The "ears" in the remnant lie along the same line as the jets seen in the X-ray image. They are probably elongated by the force of the jets from the central object. (Courtesy of S. A. Baum, R. Elston; NRAO/AUI.)

the center of the accretion disk of SS 433 is a black hole or a neutron star. Either may be able to produce the high velocities observed. Few astronomers, however, would contest the reality of the precessing jets. Very-high-resolution radio images made recently show them clearly. On images several months apart we even see the orientation of the jets change and watch conspicuous blobs of material moving outward at high speeds. The surrounding supernova remnant, W50, is reminiscent of other remnants, with a bright ring of material centered on SS 433. But the ring is distorted, with two "ears" of material sticking out along the same line as the two jets at the center. The bulge, no doubt, is somehow caused by material from the jets ramming into the old shell ejected by the original supernova; but just exactly what is going on in this bizarre system is far from fully understood. SS 433 will no doubt provide us with more surprises in the future as astronomers look at it with larger and more sensitive instruments.

A PARTIAL PICTURE

Twenty years into the study of neutron stars and pulsars, an eyeblink in the history of the universe, our knowledge of their origin and evolution is understandably incomplete. The outlook for a more complete picture, however, is more favorable than is usually the case in astronomy, for the simple reason that, in their early stages at least, pulsars age rather rapidly. If we could catch a pulsar at the moment of its creation, we could watch it fade and slow over periods of years, and we would learn much about how it generates its energy and how it affects its surroundings. Though the youngest known pulsar, the Crab, is over a thousand years old—a rather staid adult as far as pulsars go—there is no doubt that one day soon we shall be able to follow a pulsar from infancy through adolescence. As of the writing of this book, no pulsar has been detected at the position of SN 1987A, but astronomers are hoping it will not disappoint us.

Much more remains to be learned, as well, about the relation of neutron stars and black holes to supernovae and supernova remnants. What kinds of stars produce neutron stars and what kinds produce black holes? Can we see distinctive signs of a neutron star by inspecting its surrounding remnant? What percentage of neutron stars and black holes are detectable in some fashion? And how many neutron stars and black holes are out there, dark and radio silent, undetectable except for the minuscule contribution they make to the gravitational attraction of our galaxy? Our ancestors are, quite literally, among these phantoms. Long before our sun began to shine, an earlier generation of stars forged the atoms of life and, dying, scattered them to the heavens.

CHAPTER 9

Seeding the Stars

> I believe a leaf of grass is no less
> Than the journeywork of the stars
> —Whitman, *Song of Myself*

ATOMS AND THE UNIVERSE

Look around. Our familiar world is built from the debris of stars. The rocks beneath our feet, the steel and glass in our skyscrapers, the air we breathe—all are made of atoms assembled in very hot places: the interiors of stars and the outrushing shock waves of a supernova. Blown away into space, these atoms later condensed into the sun and planets of our solar system.

It may seem surprising that such mundane material can have had such a dramatic history, or, for that matter, any history at all. The elements are so much a part of our existence that we naturally assume they've always been around. Some, like gold and iron, have been known since antiquity; others, like oxygen, were first recognized several hundred years ago. Compared with the age of the universe, however, human awareness is vanishingly brief. Mark Twain once remarked that if the time since creation were represented by the height of the Eiffel Tower, the whole of human history would be no thicker than the skin of paint on the top. The advances of modern astrophysics have made us acutely sensitive to our limited temporal perspective, and we cannot take the constancy of nature for granted. Just as there was once a time when human beings had not yet evolved, so there may have been a time when the atoms themselves were not yet formed.

Greek atomist philosophers, who first conceived of atoms 2000 years ago, would have bridled at such a notion. They pictured atoms as indestructible and indivisible. (The word *atom* literally means "un-cuttable.") Although appearances might change, atoms endured for-ever. The modern viewpoint is quite different. We visualize an atom as a tiny positively charged nucleus containing protons and neutrons (collectively called *nucleons*) surrounded by a swarm of negatively charged electrons. As atoms have internal structure, it's natural to ask whether that structure was there at the beginning of time or was built up over the course of cosmic history.

One clue lies in the very simplicity of nuclear structure itself. All nuclei of a given element have the same number of protons. Hydro-gen, the lightest element, has 1; helium has 2; carbon has 6; iron has 26, and gold has 79. As a minor complication, nuclei of the same element can have different numbers of neutrons: we call these differ-ent types of nuclei *isotopes* of the element. Hydrogen, for instance, has three common isotopes: normal hydrogen with just a single pro-ton, deuterium or "heavy" hydrogen with a proton and a neutron, and tritium with a proton and two neutrons. Carbon, with 6 protons, occurs naturally in isotopes with 6, 7, or 8 neutrons in the nucleus. Iron can have 28, 30, 31, or 32. We often refer to different isotopes by giving the name of the element and the total number of nucleons (neutrons and protons) in the nucleus. Thus, the isotope of carbon with 6 neutrons is called carbon-12, and the one with 8 neutrons, carbon-14. Because protons and neutrons both have about the same weight, when we refer to carbon-12 and carbon-14, we are really distinguishing the nuclei by their weight.

It seems at least possible that the heavier nuclei were simply formed when lighter nuclei collided and combined. If the universe started with the simplest imaginable stuff, free neutrons and protons whizzing about, and if nuclei stick together when they collide, then building the elements would be a natural process that would require nothing but time.

Yet it can't be quite that simple. Nature may have constructed the nuclei by randomly assembling nucleons like blocks in a child's play-set—a process we can duplicate in particle accelerators here on Earth—but it didn't have an entirely free hand in the process. Of the infinite ways to combine protons and neutrons, only the first 92 ele-ments, consisting of several hundred distinct isotopes, have been observed in nature. Why not more? The reason is that only a limited number of nuclei are absolutely stable. Most nuclei are radioactive, spontaneously changing into other nuclei by ejecting particles or splitting into fragments. One common decay process is called beta decay, in which a neutron ejects an electron, thereby changing itself into a proton. Tritium nuclei (one proton, two neutrons) spontaneously become helium nuclei (two protons, one neutron) by the process of beta decay.

A typical tritium nucleus undergoes beta decay within 12 years after it is formed. Thus, the only tritium we see in nature is that which was produced within the last few decades; it is a by-product of nuclear power plants, for instance. The most unstable nuclei decay in tiny fractions of a second, disappearing virtually instantaneously. Others, like carbon-14, have lifetimes of several thousand years. And some, like uranium-238 (92 protons and 146 neutrons), can last for billions of years. As a general rule, among the lightest elements, like carbon, the stable isotopes have equal numbers of protons and neutrons, and among the more massive elements, like uranium, neutrons outnumber the protons by about 50%. Nuclei with too few or too many neutrons are radioactive, spontaneously decaying into other elements.

In the laboratory, short-lived isotopes can be produced and then observed in the brief time before they disappear. The elements we see in nature, however, are the ones that have survived from the time of their formation to the present. Of these, only a few are radioactive, and most of the naturally occurring radioactive ones, like uranium, have lifetimes of billions of years. The conclusion we may draw from this is that the elements formed billions of years ago. There are a few exceptions, however. Carbon-14 may be found in the air we breathe, though it decays within about 6000 years. It is produced when cosmic rays from space bombard nuclei in the upper atmosphere of Earth. Technetium-97, which lasts about 2 million years, has only been observed in the spectra of certain peculiar stars where, presumably, it is currently being formed.

Not only does nature favor stable elements, but she also seems to produce some in greater quantities than others. On Earth, the most abundant elements have a dozen or more particles in the nucleus: carbon, nitrogen, oxygen, aluminum, magnesium, silicon, and iron, for example. Light gases like helium are extremely rare. This is hardly typical of the universe at large, however. In most stars, including the sun, 9 out of every 10 atoms is a hydrogen atom; and the tenth, more than likely, is an atom of helium. The atoms that are most abundant on Earth are present only in vanishingly small numbers in the sun: 1 of every 3000 solar atoms is a carbon atom, for instance, and 1 out of every 50,000 is iron. With small variations, the solar pattern is repeated everywhere we look outside the solar system; it is a "cosmic" distribution. Hydrogen and helium, the two lightest atoms, make up virtually all of the universe. The heavier elements, though they may be essential to life on Earth, appear to be minor contaminants.

Why is Earth so overstocked in these heavy elements? The answer is that it is not overstocked at all. Rather, it has been depleted of the light elements, hydrogen and helium. Five billion years ago the sun and planets condensed out of a cloud of interstellar material, the presolar nebula. The material in the presolar nebula, like gas everywhere in the universe, was primarily hydrogen and helium. But light gases easily escape from the relatively weak gravity of Earth. The heavy elements we now see in profusion on Earth are all that was left behind after the primordial gases were driven away from the warming surface of our planet. A large planet like Jupiter, with a much stronger pull of gravity, still retains much of its original composition: it is mostly hydrogen and helium like the sun and stars.

The "cosmic" abundance pattern of the elements, ignoring local aberrations like Earth, gives us clues to how they were formed. Clearly nature favored the simplest elements, hydrogen and helium, and shunned elements like praseodymium and lutetium, which are so rare you may never have heard of them. Yet hydrogen and helium tell only part of the story, and there are revealing cosmic patterns among the heavier elements as well. Carbon, nitrogen, and oxygen (one in every few thousand atoms) are far more abundant than gold (one in every 100 billion atoms) or uranium (one in every trillion). Generally, lighter nuclei like carbon are more abundant than heavier ones like uranium, but some of the heavy elements seem unusually abundant. Iron-56 atoms outnumber potassium-39 atoms by about 200 times, for instance.

Since the late 1940s, astronomers have tried to account for the general trend in abundance patterns as well as the peculiarities. Although the details are incomplete, the following picture has emerged: The vast majority of the atoms in the universe—hydrogen and helium—were formed in the first twenty minutes of the expanding universe. The heavier elements were formed later, in stars that eventually expelled material into space, enriching the interstellar gas. The

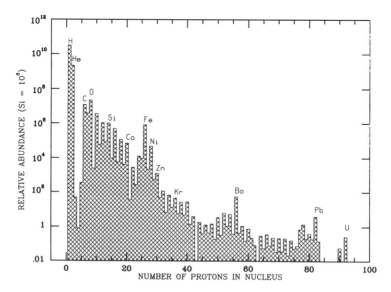

FIGURE 48: The "cosmic" abundance pattern of the elements. The graph is drawn to show the number of atoms of each element for every 10^6 silicon (Si) atoms. Hydrogen (H) and helium (He) are almost one-hundred-thousand times more abundant than silicon. Note the relatively high abundances of carbon (C), oxygen (O), nitrogen (between carbon and oxygen), and iron (Fe).

lineage of the elements, therefore, goes back to the very beginning of creation.

IN THE BEGINNING

To imagine what the universe was like long ago, we must first look at what it is like today. We know (Chapter 2) that the universe is expanding, becoming ever more spread out in space. If this has been going on for a long time, matter now separated by millions of light-years must once have been touching, packed together into an un-imaginably dense state. From the current separation and speed of the galaxies, we estimate that the expansion began about 16 billion years ago (the exact age is a matter of some debate), an event astronomers call, with hyperbolic understatement, the big bang. The tumult of a supernova pales by comparison.

The newborn universe was not only exceedingly dense but also exceedingly hot. A second after the big bang the temperature was in the billions. At these temperatures, as hot as the inside of a supernova, atoms are torn into fragments. The universe, ablaze with light, was filled with particles flying around at high speeds. As the universe expanded, however, the temperature dropped rapidly. Two to three minutes after the start of the big bang, the temperature was about ten million degrees—as high as that in the center of a star like the sun. The raw materials of atomic nuclei, protons and neutrons, still moved freely at these temperatures; nuclei might coalesce for a brief instant, but they were just as quickly torn apart.

Between 3 and about 20 minutes, the temperature and density of the entire universe were just right for cooking up stable nuclei. Protons and neutrons collided, sticking together to form isotopes of hydrogen and helium. The temperature was high enough so that collisions were relatively frequent, but low enough so that nuclei, once joined, were not immediately torn apart. At the end of this period, the universe had cooled so much that collisions no longer took place as often, and nuclear fusion effectively stopped. But by this time about 1 in every 10 atoms was an atom of helium: just about the ratio we see today throughout the universe.

The first to examine big bang element production in detail was George Gamow, a witty, freewheeling nuclear physicist who was born and educated in Russia. In 1948, he and a student, Robert Alpher, published a brief paper on the hot early universe. (Gamow, a noted punster, included Hans Bethe as a coauthor, so that future scientists would attribute the work to Alpher, Bethe, and Gamow.) The fading glow of the primal fireball, they predicted, should still be detectable today. Just as a cooling cinder radiates a dull glow, the cooling brilliance of the early universe should appear all around us. We could detect the glow as radio waves, coming at us from all directions. Nearly 20 years later, at Bell Labs in Holmdel, New Jersey, Arno Penzias and Robert Wilson discovered Gamow's radio signals, just as predicted. The discovery of the cosmic radio background, which won Penzias and Wilson a Nobel Prize, is direct evidence that our universe was once a very hot place indeed.

In 1949, together with Robert Alpher and Robert Herman, Gamow considered the proportions of the elements that should have been produced in the primal fireball. During the first 20 minutes, he found, hydrogen and helium atoms formed; but nothing else. Gamow, who had hoped to produce all the chemical elements in the big bang, was disappointed. One would think that building heavy elements would be a matter of progressively accumulating protons and neutrons. The heavier nuclei could be produced from lighter ones simply by adding a few more particles. Unfortunately, Gamow found, the process met an impassible bottleneck before it could proceed very far. The problem was that all nuclei containing five particles were wildly unstable. Add a neutron or a proton to a helium-4 nucleus and it immediately fragmented into smaller nuclei of hydrogen and helium, ending the chain of element building right there.

Was it possible to avoid the bottleneck? Not easily. The same sort of instability afflicted nuclei containing eight particles: if two helium-4 nuclei fused, the end product quickly disintegrated. Under rare conditions ten or a dozen protons and neutrons might come together swiftly enough to build up small amounts of heavy elements, but such collisions were so infrequent that they played no role in the first 20 minutes of the universe. So helium-4 was the heaviest nucleus that formed. This left an abiding mystery: where did all the other elements come from?

ELEMENT FORMATION IN STARS

Aside from the primordial fireball, the only places hot enough and dense enough for nuclear fusion are the interiors of stars and supernovae. We have already seen (Chapter 6) that stars support themselves by burning nuclei to form heavier elements, and thus it is natural to suppose that all the elements heavier than hydrogen have formed in stars over the 16 billion years since the big bang. In 1957, a decade after Gamow's seminal work, two papers, one by Alastair G. W. Cameron, and the other by Margaret and Geoffrey Burbidge, Fred Hoyle, and William Fowler, carefully considered the possibilities. By matching the relative abundances of various elements and their isotopes with the production rates of various nuclear reactions, they tried to find ways that the elements could be produced in stars. The Burbidge, Burbidge, Fowler, and Hoyle paper, a lengthy review article, earned the nickname B^2FH (after the authors' initials), and set forth the basic problems of element formation in a form that astronomers have followed ever since. Stellar nucleosynthesis, a combination of astronomy and nuclear physics, became a science in its own right.

B²FH compiled a list of nuclear reactions that synthesize the elements, but at the time they wrote they did not know for certain where all the reactions took place. Some were the regular nuclear fusion processes that occur during the lifetimes of stars (see Chapter 6). All stars, for instance, produce helium from hydrogen during the main-sequence phases of their lives. Stars of low mass (less than or equal to the sun), which constitute the large majority, never get much beyond that stage. They cannot, of course, contribute much to the abundance of heavy elements.

It's among the more massive stars, several times bulkier than the sun, that all the action takes place. They begin, like the low-mass stars, by combining hydrogen to form helium, but the path of hydrogen burning may be slightly different. If these massive stars already contain traces of heavy elements to begin with (if they are "second-generation" stars that formed from the processed material of other stars), carbon, nitrogen, and oxygen can act as catalysts to assist the fusion of hydrogen atoms. This process, called the CNO cycle, converts hydrogen to helium, just as in low-mass stars, but it also increases the concentration of nitrogen atoms, and may be a major source of the nitrogen in the universe.

Then the real business of element formation begins. After exhausting their core hydrogen and becoming red giants, massive stars assemble carbon-12 atoms from helium-4 nuclei. (Recall that this process does not occur in the primordial fireball—but only because there isn't time. The early universe cools below fusion temperatures within 20 minutes; but stars remain hot enough to burn helium for tens or hundreds of thousands of years.) The most massive stars, those more than about eight times the mass of the sun, can continue assembling nuclei, fusing helium-4 with carbon-12 to form oxygen-16, helium-4 with oxygen-16 to form neon-20, and so on until they have formed elements up to iron-56 in their cores. Then they become supernovae, spraying the heavy elements into the interstellar gas.

But this hardly exhausts the list of nuclei we see in nature. Some light nuclei seem to be skipped over in the major nuclear burning processes, whereas the nuclei heavier than iron—elements like zinc, copper, tin, and uranium—are never produced at all. Where do these elements come from? This was the major problem B²FH set out to resolve, sparking a line of research that continues to this day. Although many of the nuclear reactions that produce the heavy elements are understood, just where in the universe they occur remains something of a puzzle.

Some of the heavy elements are side effects of the major reactions that power the stars. Neutrons produced by nuclear reactions in cores of stars can collide with other nuclei there, building up some heavy elements—including technetium, strontium, zirconium, and barium—over the course of thousands of years. Because the process occurs relatively slowly, allowing the accumulating nuclei time to convert captured neutrons to protons by beta decay, B^2FH called this the s (for slow) process. Astronomers believe that the s process occurs in stars several times more massive than the sun, when they are in the late, red-giant stages of their lives. Technetium, an s-process element, is seen only in the spectra of this type of star, and because any technetium atoms formed more than a few million years ago would have disappeared by now, we suspect the s process is currently in operation there.

Among the nuclear reactions proposed by B^2FH, some seem to occur in the supernova explosion itself. A last burst of element formation occurs in the hot onionlike layers of the star, as helium nuclei quickly fuse with oxygen and silicon to build up nuclei as heavy as nickel-56. This is essentially a rapid recapitulation of the nuclear burning that occurred prior to the explosion, and astronomers therefore call it explosive oxygen and silicon burning. The radioactive nickel and cobalt produced by explosive silicon burning, as we noted earlier (Chapter 6), is a major source of energy for the early months of the supernova light curve.

And what of the heaviest elements? Those heavier than iron, especially those that have a relatively large number of neutrons compared to protons, like uranium-238, are produced almost exclusively in Type II supernovae. These neutron-rich elements result only when nuclei are subjected to an intense burst of neutrons, lasting a few minutes at most. The outgoing shock wave in a Type II supernova should generate just such a flood of neutrons as it hits atoms in its path. Bombarded by the neutrons, nuclei in the overlying gas rapidly grow fatter and fatter, and within seconds all the nuclei up to uranium can be formed. Because the neutrons are captured so rapidly compared to the slow-neutron capture process that occurs in red giants, we call this the r process. Iodine, gold, and platinum, as well as all the long-lived radioactive elements like radium and uranium were formed by the r process. When we use uranium to produce nuclear energy, we are tapping the stored energy of some ancient Type II supernova explosion.

A very few elements are not formed in stars, but in the space between stars. Atoms in the interstellar gas may, from time to time, be struck by fast-moving particles found throughout the galaxy, called cosmic rays. Cosmic rays are very likely produced in supernovae, just as Baade and Zwicky had suggested (Chapter 5). When a cosmic ray particle strikes an interstellar gas atom, the atom may split apart, forming lighter elements like lithium, beryllium, and boron. B[2]FH called this the x process.

There are other nuclear reactions, variously called the e (equilibrium) and the p (proton) processes that produce other elements. But it is not altogether clear where these exotic nuclear processes take place; the cores of red giants, the expanding shells of supernovae, and the surfaces of ordinary novae are all possibilities. Thirty years after the B[2]FH paper, astronomers still debate the details of their scenario. We are certain that the general scheme is correct: almost all the elements are formed in stars and supernovae and are scattered to the interstellar gas where, presumably, they form new generations of stars. But there are wide gaps that remain in the family tree of the elements.

SEEDING THE STARS

Nature is discreet about her origins. The working of stellar nucleosynthesis remains, for the most part, hidden in the interiors of stars. In most stars, nuclear fusion takes place so deep inside that nuclear-processed material never reaches the surface, and we never see the products of nucleosynthesis at their site of formation. Notable exceptions are a few chemically peculiar stars, like the red giants that show the presence of technetium. These stars have highly unstable atmospheres. Just as boiling water bubbles and roils, currents of rising and falling gases in these stars stir and mix together the outer layers with the energy-producing regions deep inside, lifting freshly formed elements to the surface where we see them. Stars like these, however, are rather rare.

But if the workings of stellar nucleosynthesis remain hidden, the end products do not. Many types of stars, in their later years, blow some of their outer layers back into space, returning material enriched in heavy elements to the interstellar gas. Were it not for this process of mass loss, in fact, we would not be here. It is impressive that stars produce the heavy elements; but it is far more significant that the heavy elements are returned to the pool of interstellar material from which later generations of stars can form. Dust to dust, atom to atom, the cycle of stellar life and death progresses—with a larger fraction of heavy elements in each successive generation.

Though their role is crucial, supernovae are only a part of this process. Most stars spill out some surface matter during their lifetimes without the benefit of a destructive explosion. The sun, for instance, is losing gas all the time, producing a "solar wind" that blows past Earth into the outer regions of the solar system. But only a minuscule fraction of the sun's bulk is lost through the solar wind. Later in its life, when its core hydrogen is depleted, the sun may expand to become a red giant, and then the rate of mass loss will increase. Red giants about as massive as the sun sometimes throw off shells of gas, called planetary nebulae, which look, through a telescope, like delicate rings of gas. Though the shape of a planetary nebula may resemble the X-ray image of a supernova remnant, its gases are far less energetic. A typical planetary nebula may have a diameter of half a light year, and its gases may be moving with a speed of a few tens of kilometers per second. The gases in supernova remnants, by comparison, are moving a thousand times faster.

In more massive stars, perhaps a dozen times more massive than the sun, scaled-up "stellar winds" can eject more substantial amounts of stellar material to interstellar space as the star ages. Such superwinds may strip off a large fraction of a star's outer layers, leaving little hydrogen or helium behind, and surrounding the star with a diffuse cloud of material dissipating lazily into space. We mentioned such a possibility in Chapter 6 as a prelude to the explosion of Type Ib supernovae; these are stars that seem to have lost most of their outer hydrogen before undergoing core collapse.

In the clouds of ejected gas surrounding these aging, massive stars, astronomers think they see evidence of stellar nucleosynthesis. Astronomers Nolan Walborn, Theodore Gull, and Kris Davidson, for instance, have detected extraordinarily high abundances of nitrogen in the Eta Carinae Nebula, one of the most spectacular gas clouds visible from the southern hemisphere. Eta Carinae, the central star of

FIGURE 49: A planetary nebula, NGC 7293, the Helix Nebula. (Lick Observatory photograph.)

the nebula, is a very massive star (about 100 times heavier than the sun), a few million years old, with a well-developed onionlike layering of elements inside. The high nitrogen abundance of the nebula, they suggest, is material from the hydrogen-burning shell inside the massive star, enriched by the CNO cycle and expelled into space by a high-velocity wind. Within a few thousand years, they suggest, Eta Carinae will develop a core of iron too large to sustain its own weight, and southern observers will be treated to a supernova whose brightness will exceed that of Tycho's star of 1572.

Around the Cassiopeia A supernova remnant we also see gas, in slow-moving filaments and knots, which is high in nitrogen content (see Chapter 7). This, too, may be the nuclear-enriched product of a massive star, ejected several thousand years before the supernova that formed the Cas A remnant. There are also strong indications, which we shall discuss in Chapter 11, that Supernova 1987A ejected such a nitrogen-rich shell in the millennia before it exploded.

Once a star explodes, there should be evidence of element forma-

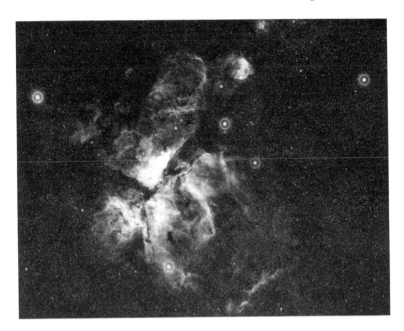

FIGURE 50: The Eta Carinae Nebula, NGC 3372. (National Optical Astronomy Observatories.)

tion in the scattered remains. The initial blast of a Type II supernova should lift off most of the gas above the iron core and hurl it into space, forming heavy elements up to uranium in the process. Among the elements ejected should be large amounts (about a tenth the mass of the sun) of radioactive nickel-56, which should decay within a few days to radioactive cobalt-56, which in turn decays within several months to form stable iron-56. Large amounts of these elements are also produced in Type I explosions. Energy produced by the radioactive cobalt probably accounts for the exponential decline of the light from Type I supernovae several months after an explosion.

There is more direct evidence of nucleosynthesis in the spectra of supernova remnants themselves. Spectra of the twisted filaments of the Crab Nebula, for instance, reveal that it is extraordinarily rich in helium. We suspect that the helium, produced by the burning of hydrogen, came from one of the outer layers of the exploding star. (The very outermost layer of the star, which should still have been rich in hydrogen, may have been ejected in a stellar wind before the supernova exploded. That is why we don't see evidence of this layer in the Crab filaments.) A number of other remnants show high abundances of oxygen, and some show high abundances of sulfur, argon, and calcium, which are produced in the late stages of a star's life.

Each element has its own particular history of formation and distribution. Several processes may form it; yet its abundance may be determined by how effectively it is scattered among the stars. Take the case of iron. In Type II supernovae the innermost core of a massive star, a sun-sized ball of solid iron, collapses to a neutron star (or perhaps a black hole). It is not expelled to space. Some iron results from the radioactive nickel and cobalt generated in the explosion itself, but only a relatively small amount is produced there. Yet iron is one of the more abundant heavy nuclei in the universe. Where does the extra iron come from?

Many astronomers would place their bets on Type I supernovae, which result (Chapter 6) from the explosive burning away of carbon-rich white dwarfs in binary systems. The carbon in a Type I supernova burns by rapidly accumulating helium nuclei to form heavy elements up to iron. In the process, the star is completely shredded and scattered into space. The remains of Type I supernovae, we might expect, would be rich in iron, and there is growing evidence that they are. Astronomers have recently detected telltale spectral lines of iron in the hollow-shell remnant of the supernova of 1006. If this is typical of Type I remnants, it lends support to our understanding of the explosive process that forms one of the most common chemical elements.

Unfortunately, the most interesting products of stellar nucleosynthesis, the rare r-process elements like gold, lead, and uranium, are so dilute, even in supernova remnants, that they cannot be detected. As a supernova remnant expands they mix with the surrounding interstellar gas, and eventually the material of the remnant becomes indistinguishable from its surroundings.

There is a cumulative effect, however, that we can detect. It is part of the great cycle of stellar life and death we have alluded to before. Stars form from interstellar gas and dust. The most massive produce heavy elements, scattering some via stellar winds, others by supernova explosions, and thereby enriching the gas from which they were born. Over the eons, the interstellar gas becomes more and more enriched by heavier elements. In regions of a galaxy where star formation and star death are more frequent, we would therefore expect to find a larger fraction of heavy elements in the interstellar gas and in the "later-generation" stars. In regions of a galaxy where only "first-generation" stars had formed from gas direct from the big bang, we should find practically no metals at all.

That is, in fact, exactly what happens. Stars in the outermost fringes of the Milky Way, for instance, particularly those in the globular clusters of stars that form a roughly spherical halo around the galaxy, are exceedingly low in heavy elements. These are, we believe, the oldest stars in the galaxy, formed from material closest in abundance to the material that emerged from the big bang. (Interestingly, heavy elements are not totally absent. We have yet to discover a "first generation" of stars, fresh from the big bang, containing no heavy elements at all.) Stars in the arms of our galaxy, where star formation is still taking place, represent a later generation of stars made from gas that was already cycled through several generations of stars. They are, accordingly, richer in heavy elements.

It is a good thing that the sun is one of those later-generation stars. Our hospitable planet—rich in the elements that constitute and support life—formed at the same time as the sun, from the same matter that made it up. Had that material been as low in metals as the most extreme cases in the halo of our galaxy, there would have been very few carbon, nitrogen, and oxygen atoms on our planet. Organic molecules would not have formed; life would not have evolved; and we would not be here to wonder about our origins.

The short account I have given here provides some idea of the complexity of the problem we face in seeking a definitive history of the elements. Clearly supernovae play a central role, both in producing and in scattering the elements. But the details of that role will remain a prime area of research for decades to come. A simple analogy may make the difficulty clear. A farm family keeps a stockpot simmering on the stove for many years. Into the pot go occasional spices, herbs, and vegetables from the garden, and occasional leftovers of game and poultry. Members of the family dip into the pot for meals, but each one adds a favorite spice or a bit of food whenever the pot runs low. On any day, the balance of ingredients in the stock reflects the history of its making. If the garden produced a lot of carrots or the father killed a goose recently, the stockpot shows it. Examining its contents, a clever detective might be able to unravel the culinary preferences of each member of the family—but the job will not be an easy one.

The pattern of element abundances in the universe is like the mix of ingredients in the stockpot. Some elements are far more common than others, and we can read the history of the elements in the peaks and valleys of the abundance curve. But just as different members of the family contribute to the overall balance of the stockpot, so different natural processes produce the elements and mix them back into the pool of material from which stars are made. Discovering which of several natural sources produced a given element is like trying to determine which family members added the carrots to the stockpot, and on what day. Our growing knowledge of supernovae—what elements they produce, and how often they explode—helps us interpret the mix of elements, and sheds light on our origins in space and time.

THE JOURNEYWORK OF THE STARS

About a hundred million supernovae have erupted in the Milky Way since it was formed. Their material, incorporated into stars and planets, is the stuff of life. But supernovae may be life-giving in another sense. They may have triggered the formation of the stars and planetary systems as well.

Stars condense from interstellar gas clouds, shrinking slowly but inexorably under their own weight. But the process must be helped along somehow, for stars seem to be forming in our galaxy far more frequently than they would be if they relied solely on the lazy drag of gravity to pull themselves together. One possible trigger might be the explosion of a nearby supernova. The outrushing shock wave from an explosion, pushing against surrounding interstellar material, could collect and compress the gas like a cosmic snowplow, speeding up the contraction of a drifting cloud, and causing it to form stars more effectively than otherwise. Our galaxy contains a few telltale signs of supernova-triggered star formation. In the constellation of Canis Minor, for instance, many newly formed stars are found right along the edge of a ring of gas that may be the remnant of an ancient supernova. The same pattern—young stars at the edge of a suspected supernova remnant—is repeated in the constellation of Cepheus and in several other places. As yet, however, the evidence is scant and circumstantial.

Nevertheless, some astronomers believe that the sun and planets themselves are by-products of a supernova that exploded about 5 billion years ago. The story of that discovery begins in 1969, on a clear winter night in Chihuahua, northern Mexico. On February 8, a ball of fire streaked across the sky accompanied by thunderous explosions, and thousands of stones, some bigger than golf balls, rained down over an area 30 miles long and several miles wide. They were fragments of a meteorite that had come apart in midflight, almost two tons of primordial stuff left over from the formation of the solar system. Meteorites such as this, which have been orbiting the sun since it first condensed from an interstellar cloud, provide a major source of information on what the solar system was like long ago.

Pieces of the Allende meteorite, named for the village of Pueblito de Allende where it fell, were snapped up by geophysicists at laboratories throughout the world. It was a time of great excitement, just prior to the first landings on the moon. Scientists were well equipped to analyze the trace elements in the meteorite, looking for clues to its origin, just as they would later study the moon rocks brought back by the Apollo astronauts. The Allende meteorite attracted particular interest because it was an example of a rare type of meteorite, called carbonaceous chondrites. Because they are older than other types of meteorites, we believe that carbonaceous chondrites are most representative of the primordial solar material.

Under a microscope, chondrites have a grainy appearance, as if they were stuck together from chips of rock and metal. Presumably these were bits of solid material that condensed from the presolar nebula—the contracting cloud that was to become the sun and planets. Among the bits and pieces in the Allende meteorite were blobs rich in calcium, aluminum, titanium, silicon, and oxygen: scientists called them calc-aluminous inclusions, or CAIs. In the early 1970s, researchers discovered abnormally high abundances of oxygen-16 and magnesium-26 in some of the CAIs in the Allende meteorite.

The extra amounts of these isotopes, argued Donald D. Clayton of Rice University, must have come from supernova material that was not thoroughly mixed together with normal interstellar material, and thus still bore direct evidence of its origin. The more revealing of the two elements was magnesium-26, a stable isotope that results from the beta decay of radioactive aluminum-26. Aluminum-26 can be formed in supernovae, but with a lifetime of about 700,000 years, it must have been trapped in the CAIs less than a million years after it was scattered into space. If a supernova went off so close to the time when the meteorites were formed—along with all the other bodies in the solar system—then perhaps that same supernova actually triggered the contraction of the presolar nebula.

Clayton's argument has sunk into the morass of controversy that tends to swallow speculative ideas about our remote origins. Subsequent researchers have questioned how chunks the size of CAIs could remain unmixed after ejection from a supernova. Some have argued that the CAI grains formed long before they were swept into the presolar nebula. Still others have tried to explain the aluminum and oxygen abundance anomalies by diffusion—the spontaneous and gradual separation of different nuclei in a solid due to their different physical properties (which would, over enough time, cause some parts of a chondrite to become depleted in an element, whereas other parts would be enhanced).

Although this controversy continues, other lines of evidence also point to a supernova origin for the solar system. Some investigators have extracted unexpectedly large amounts of xenon-129 from meteorites. Xenon is a gas formed from the radioactive decay of iodine-129, which has a lifetime of 17 million years. If the iodine had not been trapped inside the meteorite, then the xenon it produced would have gone into space, and we would not find it in the meteorite. We may conclude that these meteorites (and the solar system of which they are a part) formed within 17 million years of the supernova that created the radioactive iodine. That is a very short time on the cosmic clock.

The most recent clues are actual diamonds in the sky. A 1987 article in the journal *Nature*, for instance, reported the discovery of microscopic diamonds (only 50 angstroms in diameter) in meteorites, and suggested that they were formed by carbon compressed by the shock of a supernova explosion. The verdict is not yet in. It will be quite a while, I suspect, before we know for certain just what set the wheels of our solar system in motion, but supernovae are likely suspects.

Supernovae may have played yet another important role in the history of the solar system: they may have provided a mechanism to keep the engine of organic evolution moving along. Life, once it evolved on the surface of our planet, underwent constant variation. Different species, we believe, resulted when natural selection acted on populations with different genetic material (roughly "the survival of the fittest"). Variations appeared in a population as a result of random alterations or mutations in the DNA molecules of their reproductive cells, while environmental influences decided which variants flourished and which died out. But what produced the mutations in such a random fashion? Agents of genetic change are not hard to find: we've all heard stories of birth defects caused by chemical dumps or houses built on radioactive tailings. Environmental substances of this sort can clearly cause genetic variation. Even if you live far from a nuclear site or a uranium mine, natural radioactivity, most of it from rocks in the Earth, is constantly piercing your body, producing, from time to time, alterations in a DNA molecule here or there. Chemical and radioactive substances, because they contain heavy elements, owe their existence to supernovae.

Cosmic rays, however, which are another cause of genetic variation, can be linked directly to supernovae. First discovered in the early 1900s, cosmic rays are fast-moving protons and fragments of heavier atoms that hit Earth from all directions. They account for about one-third of all the natural radioactivity passing through your body at any moment. Some come directly from the sun, but a large number seem to come from beyond our solar system. In their 1934 paper, Baade and Zwicky suggested that cosmic rays might be produced by material ejected from supernovae. That suggestion seems quite plausible today. Supernovae go off at about the right rate and with about the right amount of energy to produce all the cosmic ray particles that we see. In addition, the abundances of various types of nuclei that we see in cosmic rays are what we would expect from supernova debris—if we make adjustments for collisions the particles might have suffered in transit between us and the supernova. It is not yet clear, however, whether the cosmic rays are simply nuclei hurled from supernovae by the force of the blast itself, or whether they are slung out at high speeds by a whirling neutron star formed afterwards. Nonetheless the link to supernovae seems well established.

If supernovae produce cosmic rays, they are, to some extent, direct agents of genetic change. Every so often a cosmic ray particle may collide with a molecule of DNA and produce a mutation in an individual organism. Add to that their indirect influence, through the formation of radioactive elements in Earth's crust, and supernovae may account for most of the genetic change that, guided by natural selection, has led from the earliest organisms to hummingbirds, humpback whales, and human beings.

It is a grand scheme, no doubt about it. The destructive power of supernovae is, paradoxically, a major agent for creation and change in the universe. Supernovae produce and distribute the elements, develop the solar system, and shape the evolution of life on one of its planets. Supernovae are at the root of our existence.

As I began this book, wandering the corridors of the Harvard Museum, I lamented the inability of astronomers to put the objects of their study on display for all to see. Writing this chapter, I recall those passages with a newfound peace of mind: everything I saw was made by the stars. Whitman was literally correct. When we consider the origins of the colorful gemstones, fossil armadillos, and stuffed lemurs that occupy the museums of the world, we behold the journeywork of the stars.

CHAPTER 10

Cosmic Dimensions

> The space of night is infinite,
> The blackness and emptiness
> Crossed only by thin bright fences
> Of logic.
> —Kenneth Rexroth, *Theory of Numbers*

A UNIVERSE OF GALAXIES

On long-exposure photographs of the heavens the galaxies crowd to-gether, one upon the other, more numerous than the stars of the Milky Way. Counting them all is as futile a task as cataloging all the stars, but we estimate that there are hundreds of billions within the limits of our largest telescopes. In each galaxy, on the average, a supernova flares once or twice a century. The implications of these simple fig-ures are surprising: every few seconds a supernova detonates some-where in the universe. Though we have not yet advanced to the point where we can detect more than a tiny fraction of these explosions, there is little doubt that our understanding of supernovae will, in time, open up new understandings of the large-scale behavior of the universe.

Supernovae play two roles in this endeavor. First, supernovae serve as tools for gauging the depths of space. Because they are so brilliant, they are visible over hundreds of millions of light-years. As we come to understand their peculiarities, astronomers hope to use supernovae to establish the dimensions of the universe with a greater precision than is now possible. Measuring these large distances, in turn, enables us to estimate how long the galaxies have been moving apart, giving us, in effect, the age of the universe.

Supernovae are more than tools; they are agents of change in the universe. Shock waves from supernova explosions are responsible for many of the spectacular patterns we see out there, from the shapes of individual galaxies to the large-scale structure of the universe. The swirling spiral arms of galaxies, according to one theory, formed when a series of supernovae exploded, one after another, like a chain of firecrackers on a single fuse. Some astronomers speculate that supernovae shaped things even larger than galaxies: clusters of galaxies tens of millions of light-years in diameter and superclusters (clusters of clusters), ten times larger again. According to these theories, even before the galaxies we see today were formed, the shock waves of the earliest generation of supernovae plowed and furrowed the primordial gas, producing galaxies in profusion and arranging them in formations of gigantic proportions.

Astrophysicists are only beginning to recognize the cosmic implications of supernovae, and hence much of what we discuss in this chapter is rather speculative. Still we find it remarkable to realize that supernovae, responsible for the creation of things as small as atoms, also had a hand in producing the largest things.

PROBING THE DEPTHS OF SPACE

The simplest questions to ask are often the hardest to answer. Among astronomers, no problem is more vexing than determining the distance to galaxies beyond the Milky Way—the extragalactic distance scale, as we call it. The problem is not purely academic. It would be nice, of course, to know precisely how far away distant galaxies are so that we can sharpen the detail on maps of the universe. More important, however, knowing the distance to the galaxies establishes the history of the cosmos. The light from distant galaxies has been traveling for eons to reach our eyes; when we look at a galaxy a billion light-years in the distance, we are also looking a billion years back into time. We see the universe as it was long ago. Reliable distances thus give us reliable timescales.

With a reliable knowledge of the extragalactic distance scale, for instance, we can determine when the universe began. Edwin Hubble, by discovering that the more distant galaxies were receding faster, introduced the notion of a universe that has been expanding from some starting point in time. Divide the distance of a galaxy by its observed speed, and you have an estimate of how long the galaxy has been traveling: the ages we calculate from a sample of well-measured galaxies range from 10 to 20 billion years. Yet which measurements are most trustworthy? If we are to take an average, which galaxies should be included in the sample? Astronomers have drawn up battle lines over the issue. Some agree with Gerard de Vaucouleurs of the University of Texas, who favors a younger universe; others agree with Allan Sandage of Mount Wilson Observatory, who argues that the universe is closer to 20 billion years than to 10. The majority are waiting to see how it all comes out.

We must measure both the speed and the distance of many galaxies to estimate the age of the universe. The Doppler shift in galaxy spectra provides a relatively precise gauge of speed. But measuring distances is a tricky business. Astronomers disagree over which of several methods to use, and how to interpret the results. Different astronomers using different methods derive different distance scales. In a game where an error of 10% is more than a billion years, the consequences of uncertainty can be profound.

It is not difficult to appreciate the difficulty. Beyond a few meters from our eyes, the world we apprehend becomes two-dimensional. We can tell what direction light is coming from, but not the distance of a source. Watch two approaching autos passing one another on a distant stretch of highway, and you will recognize the problem. Though you know one is overtaking the other, your perceptions stubbornly put them side by side. Then, just when you are sure the passing car will never make it back into its proper lane, it slides in front of the other car, with plenty of room to spare.

The difficulty is compounded in astronomy, where distance reduces many objects to indistinguishable points of light. Distance measurement, as a result, is a main concern of much of modern astronomy. For the nearest stars, it is true, astronomers can obtain relatively precise distances. If an object lies within about 100 light-years of Earth, it displays an annual back-and-forth shift, called stellar parallax (Chapter 2), resulting from our changing point of view as Earth shuttles back and forth around the sun. The nearer the star, the greater the parallax, and straightforward geometry tells us the distance. Even the nearest stars, however, show a parallax smaller than a second of arc, about the separation between a pair of auto headlights seen at a distance of 200 miles. Stars beyond 100 light-years or so, just a hair's breadth compared to the size of the universe, have parallaxes too small to measure reliably.

Beyond these nearest stars, we generally measure distances by making educated guesses of how much energy a source of light is putting out, and then comparing it with the amount of energy we receive. This was the key to many of the distance measurement schemes we outlined in Chapter 2. It was, for instance, how Edwin Hubble discovered that the Great Nebula in Andromeda, along with other spiral galaxies, lay beyond the Milky Way. He identified Cepheid variable stars in the galaxies, and knowing that they gave off an amount of light proportional to the period of their pulsations, he was able to estimate how luminous they actually were. That, combined with their measured brightness, told him they were far beyond the boundaries of the Milky Way.

Cepheid variables are among the most luminous stars, and we can spot them over several million light-years of space; but even this is a very small distance compared to the size of the universe. We can make out individual Cepheids only in about 30 of the nearest galaxies. Beyond that, as far as Cepheids are concerned we are, literally, in the dark. By studying the properties of the nearest galaxies, astronomers have hoped to find other, more conspicuous objects, and thus "bootstrap" their way out to larger and larger distances. But as the distances grow, the uncertainties grow as well.

One method is to note that the brightest stars in a galaxy all have roughly the same luminosity, as if nature set a fixed upper limit on how much power a given star could emit. Then by comparing the brightness of the brightest stars in nearby galaxies with the brightest stars in distant galaxies, one can estimate their distance. Unfortunately, the brightest stars in galaxies are not much more luminous than Cepheids. We need a more powerful source of light to probe the depths of space.

Because many stars together are brighter than one alone, why not use the total luminosity of the galaxy itself as a standard of measurement? We can, with some effort, estimate the amount of light given off by a galaxy simply by comparing it with another galaxy of the same general shape that is closer to us, and whose distance can be measured with Cepheids. Here the problem is that the sample of very nearby galaxies includes very few objects, and that galaxies of differ-

FIGURE 51: Supernova 1961I in the spiral galaxy NGC 4303. (Lick Observatory photograph.)

ent shapes and sizes may have different luminosities, making estimates of the "typical" luminosity very uncertain.

Supernovae may be the bright light we are looking for. Astronomers have long recognized that Type I supernovae are an exceedingly uniform lot, and that at maximum they all are equally powerful—about 10 billion times more luminous than the sun. This is more than 10 thousand times brighter than the most luminous normal stars in galaxies, and thus supernovae should be clearly visible for hundreds of millions of light-years, standing out above the fainter glow of their parent galaxies. Type I supernovae may thus serve us well as "standard candles": the apparent brightness of a Type I at maximum would be a direct measure of its distance. With his usual prescience, Fritz Zwicky long ago championed using Type I supernovae to establish the distance scale, though, as was often the case, he represented his case too strongly. At a recent symposium on the subject, Allan Sandage remarked that Zwicky's "acerbic (some would even say scabrous) rejection of all other methods" did nothing to win converts. Now, as the study of supernovae matures, Zwicky's program is being revived.

Type I supernovae are useful because their behavior is easily predictable. Type II are not. Each seems to behave differently from most of its predecessors, and astronomers cannot tell, when looking at a Type II, just how much light is being given off at any given time. Actually, there is probably some variation in the maximum light emitted from one Type I to another, but the variation is less extreme. As we begin to understand both types of supernovae better, however, we will recognize more of the distinguishing characteristics of individual cases, and they will yield more and more precise values of distance. One promising avenue of research was initiated by David Branch, of the University of Oklahoma, who has developed reliable computer programs that enable us to reproduce the entire spectrum of light emitted by a supernova. Using such programs to reproduce the spectrum we see in an individual case, we can predict the light output of a specific supernova and thus determine how far away it is. This case-by-case method may, in the long run, be even more precise than simply assuming that all Type I supernovae are close to some average luminosity at maximum. Using appropriate computer models we may eventually be able to apply the method to Type II supernovae as well.

Norbert Bartel and co-workers at the Harvard–Smithsonian Center for Astrophysics have pioneered another way of using supernovae as distance indicators. They look at the radio emission from the expanding shell of a supernova. In ordinary optical telescopes, the expanding shell cannot be distinguished from an infinitesimal dot. Bartel, however, uses very-long-baseline interferometry, a technique in which radio telescopes separated by thousands of kilometers look at the same object. The signals collected by each telescope are combined at a central location, and with much computer processing, they yield images of striking resolution and clarity, giving details thousands of times smaller than a single telescope can. In the case of a 1979 supernova that appeared in the galaxy M100, Bartel and his colleagues were able to measure the angular size of the expanding shell quite effectively. (It was less than a thousandth of a second of arc in diameter, roughly the size of a dime seen from a distance of 1000 miles.)

They could also calculate the size of the shell in kilometers from the speed of the gases measured by the Doppler shifts in the optical spectra. Straightforward trigonometry then gave them the distance of the supernova. (This is essentially the method used to measure the distance of Nova Persei 1901, which I described in Chapter 5.) Bartel has since applied the method to another supernova. As it becomes more refined, radio observations may produce distances that are as reliable, or better, than any obtained by more conventional methods.

In both these methods, supernovae are merely used as bigger and more powerful standard candles. For more accurate distance determinations, we need a deeper understanding of just how much light they produce and how fast they expand. We don't have that understanding yet. At present neither Type I supernovae nor radio measurements yield distances of galaxies or ages of the universe that are any better than those we already have from other techniques: supernovae can only tell us that the universe is greater than 10 and less than 20 billion years old. With time and research, however, supernovae may eventually help us set the date of creation more precisely: if not to the hour, at least to the nearest billion years. At present, astronomers would be quite happy with that degree of certainty.

SHAPING THE GALAXIES

When we see a meteor flash through the nighttime sky, we regard it as a minor disturbance, a chance encounter with a bit of space junk. Yet Earth and the other planets probably formed by the accumulation of large amounts of such debris—the meteor is just a part of the ongoing process of planetary formation, and Earth still accumulates several hundred tons of meteoric material each day. Similarly, when we see a supernova flare in a distant spiral galaxy, it is not just another bit of celestial pyrotechnics. Supernovae, we believe, shape the spirals, creating the exquisite swirls of light that draw our attention in photographs of deep space. When we see a supernova, we are watching the hammer blows that sculpt the galaxies.

The spiral galaxies are only one of three types of galaxies recognized by Edwin Hubble. There are elliptical galaxies, which appear spherical or egglike, and irregular galaxies, which appear as rather shapeless blobs. But the spirals surely rank among the most striking forms of nature, and when most people think of galaxies, they think of spirals. Most are like our own Milky Way, flattened disks of stars, 100,000 light-years across, with tens to hundreds of billions of stars whirling around a central bulge, much as the planets circle the sun. Looking at a spiral galaxy from afar we see wispy streamers of stars (astronomers call them spiral arms), curving out from the center as if they were sprayed from some fireworks display.

All this, in a sense, is window dressing. Photographs of spiral galaxies accentuate the most luminous objects in a galaxy: massive blue stars, newly formed and very hot, along with huge glowing clouds of gas illuminated by the blue stars. The blue stars and gas trace out the spiral arms. But between the arms are many faint red stars, along with clouds of cool gas, which escape our immediate attention simply because they don't emit a lot of light. When all is accounted for, the spiral arms are just regions of enhanced light embedded in an uninterrupted disk of material. They seem more like an ornamental flourish than a fundamental structural feature of galaxies.

FIGURE 52: An Sb galaxy, M81. (National Optical Astronomy Observatories.)

Yet superficial or not, how can we account for the lovely shapes we see? Hubble tried to attack the problem of spiral form the way early biologists like Carl Linnaeus approached the diversity of species, by first classifying and establishing a sequence of shapes. Spirals, he found, showed a marked variation in how prominent the central bulge was and how tightly wound the arms appeared. He gave the name of type Sa galaxies to those with a prominent bulge and tightly wound arms, almost indistinguishable as separate objects. Type Sb galaxies were looser. Type Sc galaxies, at the other extreme, had a very small central bulge and sprawling spiral arms. It was possible that all spirals diverged from a common origin, like apes and humans. Or perhaps they represented different stages in the development of a single type; perhaps Sa's developed into Sb's and Sc's or vice versa.

The motions of stars in galaxies suggested, in fact, that the arms of galaxies might wind up in time, causing Sc's to evolve into Sa's. If we were to watch a time-lapse motion picture of a galaxy spinning, we would see that each star in a galaxy moves around the center like a planet in a solar system. The stars closer to the center of a galaxy complete their orbits in less time than the stars farther out. (To get some idea of the timescale, the sun, about three-fifths of the way out from the center of our galaxy, goes around once every 250 million years.) The swifter motion of stars closer in is called the differential rotation of the galaxy.

Differential rotation should cause the arms of galaxies to wind up. Imagine, for instance, a group of stars arranged in a straight line pointing directly away from the center of a galaxy. Now watch the line of stars as the galaxy rotates. Because the innermost stars orbit faster, they leave the outermost stars trailing behind, and the line bends initially into a spiral shape. Within a few rotations the line winds up from the inside out, like a tightly coiled watch spring. Eventually it is torn to pieces and dissipates. This is just what we might expect if Sc galaxies evolve into Sa galaxies.

The problem remains that there is no other evidence that Sc galaxies are any younger than Sa galaxies. When we look at galaxies at great distances we are looking further into the past, at a time when the galaxies were younger. If Sc galaxies wind up to form Sa galaxies, we might expect to find more Sc's among the younger objects, at larger distances from us. We don't. Moreover, at the rate galaxies rotate, we would expect to find most of them wound up tightly by

FIGURE 53: An Sc galaxy, M33. (Lick Observatory photograph.)

now, yet there are many Sc and Sb galaxies close to us. Our own Milky Way, which formed 10 to 15 billion years ago, is a relatively loosely wound spiral. How can galaxies still show spiral patterns after billions of years of differential rotation? Astronomers have come to call this contradiction between the spiral patterns of the galaxies and the motions of the stars that make them up the "winding dilemma."

There are two possibilities. Something may sustain the spiral, preserving the pattern as the stars move through it. We see such things happen on a smaller scale all the time. A jutting rock can sustain chevrons of white water in a stream: though new water constantly passes through the foam, the same pattern remains. Alternatively, something may constantly regenerate new arms, forming spiral patterns of bright stars as old ones are wound up and disappear. Both mechanisms, apparently, play a part in patterning the spirals; in some galaxies, they are sustained, in others, continuously regenerated.

Let us first look at how spiral patterns might be sustained. Stars typically circle around the galaxy once every hundred million years or so. They spend only a small part of that time, a few tens of millions of years or so, in a given spiral arm. Because the bright, massive stars that mark the spiral arms typically live only a few tens of millions of years, they must be formed in the spiral arms. Otherwise we would see them in the intervening spaces as well.

Nature forms the bright, massive stars from interstellar clouds of gas by a process of contraction and heating that I described in an earlier chapter. The gas clouds, like the stars, orbit around the center of the galaxy, and only spend a short time passing through the spiral arms. The trick is to get the gas clouds to collapse and form stars in the spiral arms and nowhere else, producing a concentration of bright stars that marks the spiral. In the late 1950s and early 1960s Bertil Lindblad, a Swedish astronomer, proposed that a sort of galactic traffic jam, called a density wave, could produce a bunching up of the clouds of gas, triggering rapid star formation. Later work by Chia-Chiao Lin and Frank Shu of MIT showed that such density waves could indeed take on the spiral patterns we see.

According to the density wave theory, a galaxy may form as a disk without spiral arms. Some disturbance, such as the gravitational nudge from a passing galaxy, causes a ripple in the gas of a galaxy, not unlike that caused by a wind blowing over a quiet pond. The ripple takes on a symmetric geometric shape, just as a drumhead, when struck, vibrates in a predictable, symmetric pattern. In this case the ripple in the galaxy takes on the form of a symmetric spiral. Stars and gas orbiting the galaxy pass into the ripple and are bunched up, and thus exert a greater-than-average gravitational pull on their surroundings. The bunched-up gas collapses under its own weight, fragments, and forms stars. All the newborn stars continue to orbit, and eventually they pass out of the density enhancement. But the most massive, hottest stars die out within a few million years—before they have had a chance to move very far from their place of birth. Thus, in the density wave theory, hot blue stars mark out the density waves themselves. The spiral pattern persists, though the stars that delineate it are constantly changing.

Supernovae are only incidental to the density wave theory. The bright stars produced by the density waves are the same massive stars that become Type II supernovae as they die out, so if density waves produce spiral structure, then Type II supernovae should occur preferentially along the spiral arms.

But supernovae are central to the other theory of spiral structure, in which the spirals are constantly being regenerated anew. The theo-

FIGURE 54: A "grand design" spiral: NGC 628, type Sc. Photographed in blue light to emphasize hot young stars. (Courtesy of Debra Meloy Elmegreen and Bruce Elmegreen.)

ry is based on an idea proposed in 1976 by Ivan Mueller of Ohio State University and David Arnett of the University of Chicago. Arnett and Mueller had noted that if supernova explosions go off in a region rich in interstellar gas, the shock from the explosion can compress gas ahead of it, triggering new star formation, which in turn produces further supernovae, which trigger more star formation, and so on. Thus, the relatively local effect of a supernova can have far-reaching consequences. "Sequential" star formation, in which supernovae spread new stars like a sneeze spreads the flu, was taken up by a number of other astronomers and is today a much discussed, though still speculative, theory.

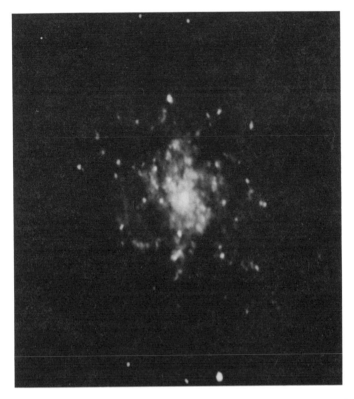

FIGURE 55: A "flocculent" or "chaotic" spiral: NGC 7793, type Sc. Photographed in blue light to emphasize hot young stars. (Courtesy of Debra Meloy Elmegreen and Bruce Elmegreen.)

In 1978, Philip Seiden and Humberto Gerola of IBM's Thomas J. Watson Research Center took this notion and applied it to a galaxy of rotating stars. Suppose a supernova goes off at random in a galaxy. It triggers the formation of new stars nearby, some of which soon explode (within a million years or so), forming still more new stars. In a relatively short time, a chain of bright stars is formed. The stars closer to the center of the galaxy move faster, however, and the chain is distorted into a spiral shape, just as we noted earlier. After a few revolutions, the line of stars has been dissipated, and the spiral is no longer visible, but by then other supernovae have gone off randomly in the galaxy, producing new chains of stars, and new spiral arms. In Seiden and Gerola's theory, called stochastic (i.e., random) self-propagating star formation, the spiral structure is always disappearing and always being regenerated. Using computer simulations, they have produced striking motion pictures showing the evolution of the spiral structure of a galaxy over billions of years. Here and there a supernova flares, patches of bright stars form, align themselves in spiral shapes, and wind into oblivion. New stars are formed as old ones disappear. The mechanism is surprisingly effective, considering how simple the idea is.

In all probability, both density waves and supernova-induced star formation play a role in producing and sustaining spiral structure. Some spirals, called "grand design" spirals, like NGC 628 pictured in Figure 54, show a symmetry and sharpness of structure that suggests that density waves are the major influence. Others, called "flocculent" or "chaotic" spirals, like NGC 7793 (Figure 55), show a patchiness that suggests stochastic star formation induced by supernovae may be more important. No astronomer would claim that supernovae are the sole creative force in the cosmos. Yet they have left an unmistakable mark on the shapes of the spiral galaxies.

PRIMORDIAL SUPERNOVAE

One of the most remarkable discoveries of modern astrophysics is the structure of the universe on the largest scale. I am speaking here of a scale far larger than any we have spoken of in our discussion of galaxies. The galaxies, which are admittedly enormous aggregations of stars, seem to be bunched into even larger groups—clusters of galaxies—containing anywhere from a few dozen to several thousand members. Whereas a single spiral galaxy may span 100,000 light-years from side to side, a cluster of galaxies may be 50 times larger, spanning 5 million light-years or more. Our own Milky Way is part of a relatively small assemblage, called the Local Group, which also includes the Magellanic Clouds, M33, and the Great Nebula in Andromeda.

In the last decade, astronomers have mapped even larger structures called superclusters of galaxies that span hundreds of millions of light-years. A vague three-dimensional outline of the cosmos is now emerging. But it's not what we might have expected. Seen on this largest scale, the distribution of matter in the universe seems exceedingly patchy. The superclusters are long strings and filaments of galaxies with vast spaces in between them. In the direction of the constellation of Bootes, for instance, there is a giant void, a region 150 million light-years in diameter, that seems to contain virtually nothing (or at least nothing that emits light). These large voids seem scattered everywhere we look, separating one supercluster from another.

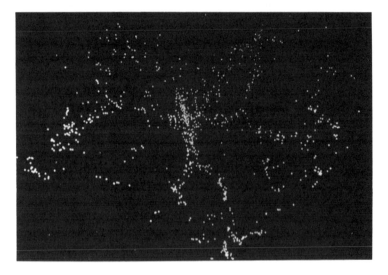

FIGURE 56: The "bubbly" structure of the universe is evident on this plot of the spatial distribution of 1100 galaxies in the direction of the Coma Cluster as obtained by Valérie de Lapparent, Margaret J. Geller, and John P. Huchra of the Harvard–Smithsonian Center for Astrophysics. (Courtesy of Margaret Geller and Michael Kurtz.)

Some visualize the structure of the universe as a sponge, with large holes separated by shells of galaxies. Others have likened the superclusters to the skins of giant bubbles. The universe appears foamy, they say. It is mostly empty space with a thin suds made of many luminous galaxies.

Scientists find the bubbly nature of the universe surprising because it suggests that something other than mere chance decided when and where the galaxies were to form. Before they knew what the large-scale structure of the universe was like, most astronomers would have bet that a map of the universe might resemble the pattern of raindrops on an asphalt parking lot: random dots with no large blank areas left untouched. Because galaxies are formed when matter collapses under its own weight, astronomers reasoned, one might expect them to form randomly wherever the primordial gas from the big bang was a little more compressed than usual. The current observations show that the distribution of galaxies is not random at all. They are found at the well-defined edges of giant empty regions. What formed them there? Why is there so much empty space between the superclusters?

No one knows for sure, but one particularly provocative idea was raised in 1981 by Princeton astrophysicists Jeremiah Ostriker and Lennox Cowie, and independently by Satoru Ikeuchi of Hokkaido University in Japan. Ostriker and Cowie proposed that the large-scale structure was impressed on the universe by supernova explosions that took place during the first billion years or so after the big bang, at a time when the galaxies we see today had not yet formed.

The universe at this time consisted mostly of hydrogen gas, expanding because of the big bang. Yet there were a few galaxies as well, formed here and there by denser, randomly distributed concentrations of gas. In the hearts of these firstborn galaxies, there were billions of massive stars (more than we find in a typical galaxy today), which burned their nuclear fuel at exceedingly rapid rates. Because there were so many massive stars in these early galaxies, there were also many supernovae. During a span of perhaps 100 million years, a billion supernovae might have gone off in a galaxy—a rate at least 1000 times greater than that we see today. The expanding shock waves from these many supernovae blasted through the gas of the universe, clearing out holes, and pushing shells of gas outward like gigantic snowplows. New galaxies formed in the plowed-up gas around the edge of the shells. They contained new massive stars that became supernovae and blew out additional bubbles in the surrounding gas, forming new galaxies in turn. The entire process resembles the scenario for sequential star formation in a single galaxy, but on a scale ten thousand times larger. By the time the current generation of galaxies (including our own) had formed around 10 or 15 billion years ago, the old supergalaxies were gone. All that remained of the early epoch of furious galaxy formation and equally furious star death was the bubblelike distribution of the superclusters.

Ostriker and Cowie's theory is attractive in many ways. It neatly explains the cosmic sponginess of the universe. At the same time it helps clear up a seemingly unrelated question: why can we not find a single star in our galaxy that contains only hydrogen and helium atoms? As we noted in Chapter 9, no atoms heavier than helium (except a tiny bit of lithium) should have been formed in the big bang. Yet even the oldest stars in the Milky Way, the stars in the outermost halo of the galaxy, show significant amounts of carbon, nitrogen, iron, and other elements. If Ostriker and Cowie are correct, however, these elements were first formed in the primordial supernovae that excavated the interstellar voids and shaped the superclusters. The first round of heavy element formation took place before our galaxy was formed.

Clever though it is, Ostriker and Cowie's work is still rank speculation, and there are many who question its applicability. Can so many supernovae explode in close enough synchronization to scoop out cosmic-sized voids? Even if so many explosions were to take place, would the structure we see necessarily result? Nothing approaching this explosive scale exists for comparison in the universe today. And can't we devise alternative theories that can account for the formation of superclusters and voids without the need for explosions?

Questions of the origin of the universe are always among the most controversial in astronomy, and we can expect no quick resolution of the debate. Whatever its outcome, the theory of primordial supernovae highlights how interconnected the various domains of astrophysics have become. The nuclei of atoms may bear the stamp of the force that formed the galaxies.

CHAPTER 11

A Death in the Neighborhood

> And all about the cosmic sky,
>> The black that lies beyond our blue,
> Dead stars innumerable lie,
>> And stars of red and angry hue
>> Not dead but doomed to die.
> —Julian Huxley, *Cosmic Death*

A DOT IN A CLOUD OF LIGHT

The night of Monday, February 23, 1987 settled clear and cloudless over the coastal mountains of northern Chile. Here, along the western edge of the Andes, the land is arid and parched, for the prevailing easterly winds drop their moisture as they ascend steep slopes far to the east, feeding the great rivers of Argentina and Uruguay. The coastal mountains support little indigenous life, but they are ideal for astronomy. Astronomers, seeking crystal skies and a view of the southern heavens not visible from the northern hemisphere, have established several observatories on barren mountaintops not far from the coastal city of La Serena. Of these, Las Campanas Observatory, operated by the Carnegie Institution of Washington, is perhaps the most isolated. With only four telescopes on the mountain, there are rarely more than a handful of astronomers and technicians there on any given night.

Oscar Duhalde, night assistant at the Carnegie 40-inch telescope on Las Campanas, was working with astronomers Barry Madore of the University of Toronto and Robert Jedrzejewski of the Space Telescope Science Institute in Baltimore on the evening of the 23rd. Night assistants are to astronomers what guides are to fishermen. They are usually more familiar with the complex operations of their telescope than an astronomer, who may use it only a few days each year. The night assistant pilots the telescope, pointing it to designated spots in the sky and making sure that all the equipment operates properly.

February 23rd was a routine evening. Madore and Jedrzejewski would ask for a star, Duhalde would set up the telescope and then wait while the sensitive solid-state camera attached to it made an exposure. He passed the time by listening to a tape by the Talking Heads. At about midnight, Duhalde went to the adjacent kitchen to make himself a cup of coffee. While the water was heating, he stepped outside to check the clarity of the sky. The doorway faced south. There, he knew, was the Large Magellanic Cloud, one of the nearest galaxies, a familiar and reassuring glow. At one edge of the LMC would be a brighter patch of light, the 30 Doradus Nebula, also known as the Tarantula Nebula, a large hot cloud of hydrogen gas illuminated by some very massive stars. Duhalde knew that if 30

FIGURE 57: Oscar Duhalde.

Doradus was hard to see, there was a slight haze overhead; but if the nebula appeared bright and clear, it was a very good night. Looking south, Duhalde could see the nebula with no difficulty at all. But to his surprise, just to the southwest of the nebula, he saw a distinct dot of light—a new star that had not been there on any of the many times he had looked before. Duhalde went back inside, got his coffee, and went to tell the astronomers what he had seen.

Though he didn't realize it until later, Duhalde was the first person to see a supernova with the unaided eye in 383 years. He was curious about the object, however, for he'd never seen anything like it before, and he knew something odd was going on. But when he returned to the control room, it was time to move the telescope to another star, and in the hustle of the moment, Duhalde forgot to mention his discovery. "We must have been working him too hard," recalled Jedrzejewski later, half humorously. Yet there was no urgent reason to discuss the sky. To the astronomers, it was just a fine evening, a time to observe as many stars as time permitted. The work went on, uninterrupted, for several hours.

Just a few minutes before Duhalde's sighting and a few hundred yards from where he had been standing, Ian Shelton, resident observer at the University of Toronto's modest 24-inch telescope on Las Campanas, began a long exposure of the Large Magellanic Cloud. Lean and intense, with a deep interest in astronomy and photography, Shelton had lived on the top of Las Campanas for 4 of his 29 years. He made observations for astronomers at the University of Toronto, and in his spare time, he made observations for himself.

Shelton had persuaded Bill Kunkel, the administrator of Las Campanas, to let him use a 10-inch photographic telescope on the mountain to patrol the Large Magellanic Cloud for variable stars and common novae. It was a balky old instrument, housed in a shed with a roof that rolled off by hand. To get the sharpest images, Shelton had to guide it by eye as it tracked the stars, observing a reference star through an eyepiece attached to the side of the telescope and jogging the drive motors slightly from time to time as it drifted away from the object of interest. Shelton would jockey the telescope for three hours at a stretch. Guiding was tedious work, but he found it satisfying to see the needle-sharp star images on the glass photographic plates after a good night of observing. A good plate was a sign of a keen eye and a steady hand on the control buttons of the old telescope.

February 23rd was only the third night of Shelton's patrol of the

FIGURE 58: Ian Shelton. The small telescope with which he discovered SN 1987A can be seen behind him. (Courtesy of Robert Kirshner.)

Large Magellanic Cloud. The first two nights had not gone particularly well—the star images were not as sharp as they could have been—and Shelton wanted to get at least one flawless plate to begin with. He began an exposure just before midnight, and by 2:30 a.m., when a gust of wind blew the shed roof shut, he called it quits. Though the hour was late, Shelton took the plate to the darkroom, developed it, and snapped on a light to see how he'd done.

What he saw was encouraging: star images were sharp from one edge of the plate to the other. Then Shelton started; for a moment he thought he had fouled up again. Not far from the spiderlike shape of the 30 Doradus Nebula was a large dark spot. Was this a "plate flaw," a defect that had been there all along? Had the plate somehow been scratched or inadvertently exposed to light? After a few minutes Shelton convinced himself that the image was real; it was a bright star, about fifth magnitude, far brighter than anything else in the Large Magellanic Cloud. He compared it with a plate from the night before: there was nothing brighter than 12th magnitude at the spot, 600 times fainter than the new star. (The most luminous stars in the Large Magellanic Cloud appear no brighter than 12th magnitude.) Only then did he run outside and look. The dot of light was just where he expected it to be, faint to his unaided eye, but unmistakable. It was a nova, most likely, right in the heart of the Large Magellanic Cloud.

Shelton walked quickly into the control room of the nearby 40-

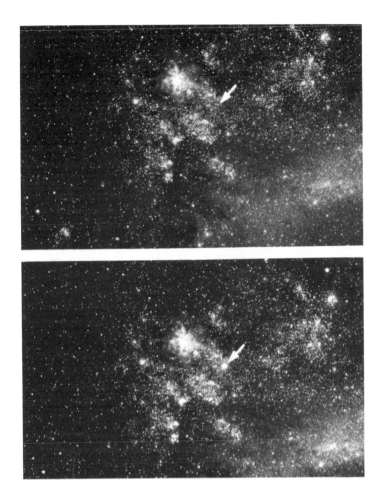

FIGURE 59: (a) Prediscovery photograph and (b) discovery photograph of SN 1987A made by Ian Shelton. (Courtesy of Robert Garrison, University of Toronto.)

inch telescope with the news. He thought he'd seen a nova in the Large Magellanic Cloud. Duhalde remembered and said, yes, he too had seen it earlier. But was it surely a nova? At the distance of the Cloud, even a very luminous common nova would be no brighter than 8th magnitude, too faint to be seen without a telescope. They all went out to look. The night was still exceptionally clear, and the new star stood out persistently against the hazy glow of the Large Magellanic Cloud. At fifth magnitude, it didn't knock them over by any means. Nevertheless, Barry Madore assured them, if the object they were seeing was at the distance of the Large Magellanic Cloud, about 160,000 light-years, it was too bright to be an ordinary nova. Shelton had photographed a supernova on the rise. If it followed true to form, it might be the brightest one since Kepler's.

The rush began. They had to spread the news immediately so that others could observe the supernova while it was still rising to maximum brightness. Moreover, if Shelton and Duhalde had been the first to discover it, they had to establish their priority. It was already growing dark in New Zealand and Australia; someone there would surely spot the supernova and claim the credit for the discovery.

The established procedure was to notify Brian Marsden at the Central Bureau for Astronomical Telegrams in Cambridge, Massachusetts. "No tree falls in a forest, unless Brian says it falls," according to Harvard astronomer Robert Kirshner. Marsden's job is to receive reports of astronomical discoveries and important celestial events—novae, comets, asteroids, and the like—and, after verifying the report is valid, notify the astronomical community as quickly as possible. Several times a week, whenever something notable occurs, Marsden's office sends out telegrams and postcards to astronomers and observatories throughout the world. Marsden also maintains a list of comets, asteroids, and supernova discoveries.

The supernova was sinking toward the southwest horizon as Barry Madore placed one long-distance call after another to the Bureau. In the dead of night (it is two hours earlier in Cambridge than in Chile), Marsden's phone rang unanswered. Finally the four excited observers gave a message to another night assistant, Angel Guerra, who was driving down that morning to La Serena, 100 miles to the south. Guerra passed the message on to his boss, Bill Kunkel, who telexed it to Cambridge. Shown is a reproduction of the original telex as Marsden received it around 9 a.m. that day.

FIGURE 60: Brian Marsden in the office of the Central Bureau for Astronomical Telegrams. (Courtesy of Frederic Marschall.)

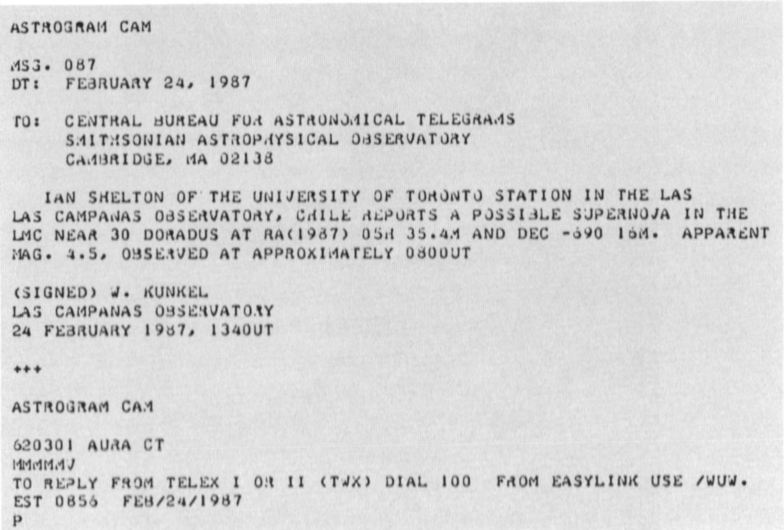

```
ASTROGRAM CAM

MSG. 087
DT:   FEBRUARY 24, 1987

TO:   CENTRAL BUREAU FOR ASTRONOMICAL TELEGRAMS
      SMITHSONIAN ASTROPHYSICAL OBSERVATORY
      CAMBRIDGE, MA 02138

      IAN SHELTON OF THE UNIVERSITY OF TORONTO STATION IN THE LAS
LAS CAMPANAS OBSERVATORY, CHILE REPORTS A POSSIBLE SUPERNOVA IN THE
LMC NEAR 30 DORADUS AT RA(1987) 05H 35.4M AND DEC -690 16M.  APPARENT
MAG. 4.5, OBSERVED AT APPROXIMATELY 0800UT

(SIGNED) W. KUNKEL
LAS CAMPANAS OBSERVATORY
24 FEBRUARY 1987, 1340UT

+++

ASTROGRAM CAM

620301 AURA CT
MMMMMJ
TO REPLY FROM TELEX I OR II (TWX) DIAL 100   FROM EASYLINK USE /WUW.
EST 0856  FEB/24/1987
P
```

FIGURE 61: Telex announcing Shelton's discovery of SN 1987A. (Courtesy of Brian Marsden.)

Reports of the supernova began to flood into Marsden's crowded Cambridge office even before the official telegrams could be sent out. From Nelson, New Zealand came word that Albert Jones, an amateur who monitored variable stars and novae, had spotted the supernova through scattered clouds about four hours after Shelton. Over a span of several hours it had brightened noticeably, according to Jones. The supernova was undeniably undergoing its initial rise to maximum brilliance.

Night swept westward, revealing the brightening star to a growing number of astronomers. One ironic note came from R. H. McNaught, an observer at Siding Spring Observatory in Australia,

FIGURE 62: Albert Jones. (*Nelson Evening Mail.*)

who had photographed the supernova several times that evening, beginning shortly after Jones had sighted it. McNaught, using a Schmidt camera on a regular patrol of the sky, had actually photographed the supernova the night *before*, shining at 6th magnitude (barely visible to the naked eye). But he'd delayed developing and examining the film, and thus missed being the first to realize the supernova had gone off. The entire astronomical community, in addition, might have profited from the chance to observe the supernova at an earlier stage. McNaught's February 23rd photograph, nonetheless, is the earliest showing the supernova on the rise, and enables us to trace the light curve of the supernova back almost to the time of detonation.

By 10 a.m. Eastern Standard Time on the 24th, Marsden had composed a notice to send out—the first of many in the hectic weeks that followed. According to accepted practice, supernovae are designated by year and by capital letters of the alphabet assigned in the order they are discovered. This was the first of the new year. "SUPERNOVA 1987A IN THE LARGE MAGELLANIC CLOUD" read the first line. Shelton, Duhalde, and Jones were credited with the discovery. As the telegrams were sent, news spread swiftly through the grapevine of astronomers. The supernova of a lifetime—of several dozen lifetimes—had arrived.

WHICH STAR EXPLODED?

On the morning of the 24th, Robert Kirshner of Harvard University, who has been studying supernovae and their remnants for nearly two decades, received a phone call from a friend, J. Craig Wheeler at the University of Texas. One of Wheeler's graduate students, visiting Toronto, had heard of the supernova in the Large Magellanic Cloud, and Wheeler knew that Kirshner would want to know right away. For several years, Kirshner had a standing "target of opportunity" proposal for observing supernovae using the International Ultraviolet Explorer (IUE), a satellite-mounted telescope that takes spectra at ultraviolet wavelengths blocked by Earth's atmosphere. Observations on the satellite are usually scheduled months in advance, but Kirshner's proposal entitled him to preempt the telescope in the event that a bright supernova appeared. He'd used the IUE on short notice to observe a supernova in the galaxy NGC 5128 in May

1986. During the first days of a supernova, when it is at its hottest, most of the light is emitted at ultraviolet wavelengths; but later, the ultraviolet intensity drops markedly. Early observation, therefore, was essential.

As he reached across his desk to call the IUE, the phone rang. It was the IUE office, informing him of the supernova, and asking him what arrangements he wanted to make for observing. Kirshner hurried to NASA's Goddard Space Flight Center, just outside of Washington, D.C., where the satellite control room is located. By midafternoon, Kirshner and a collaborator from Goddard, George Sonneborn, had begun to take spectra of SN 1987A.

Throughout the world, the reaction was the same. Phones rang; hallways filled with astronomers busily discussing the fast-breaking developments and making plans to study the supernova themselves. At major observatories in the southern hemisphere, observing schedules were altered to fit in time for the new discovery; observations were carried out with every conceivable type of instrument. In Argen-

FIGURE 63: Robert Kirshner in his office. (Courtesy of Frederic Marschall.)

tina, Chile, Australia, New Zealand, and South Africa, telescopes began to monitor the rising brightness and changing spectrum of the supernova. Some of the equipment was hastily modified to reduce the brilliance of the supernova: many modern astronomical instruments are designed to work only with objects hundreds or thousands of times fainter than SN 1987A. Astronomers, who usually strain for more light, were presented with an overabundance.

There was an overabundance of data as well. Swamped by reports, Marsden's office was working full-time to keep astronomers abreast of the latest developments. There were hundreds of estimates of the brightness and color of the supernova; there were descriptions of the changing spectrum; there were reports of a fleeting burst of radio waves from the supernova. And there were negative reports as well: no X rays had been detected by a recently launched Japanese satellite called Ginga ("Galaxy"). The supernova, so bright at optical wavelengths, was not yet emitting enough X rays to be detected. Within two weeks, Marsden and his assistants sent out 30 telegrams, a twice-daily chronicle of discovery and excitement.

Astronomers found it remarkable to be able to examine a supernova in such exquisite detail. And it was remarkable to watch it change so rapidly. When Kirshner and Sonneborn began observing, SN 1987A was quite strong in the ultraviolet, but this was clearly beginning to fade. As the ultraviolet dimmed, however, observers at optical wavelengths noted a brightening. This was to be expected. The expanding supernova should cool rapidly during the first days, from surface temperatures of hundreds of thousands of degrees to only ten or fifteen thousand degrees. Its color should change, just as a bar of steel goes from white to yellow to red as it cools. At its hottest, when the shock burst through the surface of the star, it would emit primarily short-wavelength ultraviolet radiation. But as it cooled, more and more radiation would appear at visible wavelengths, just as optical astronomers saw in the first few days.

In spectra of the supernova, astronomers could see strong lines of hydrogen, indicating that it was likely a Type II blast. The lines showed Doppler shifts of gas moving outward at speeds as high as 40,000 kilometers per second, more than 10% of the speed of light. After a few days, the Doppler shifts decreased. This did not indicate that the gas was slowing down, as one might normally think; it takes thousands of years before the expanding gases from a supernova slow appreciably. The decreasing Doppler shifts appeared because the outer layers of the supernova, which move faster, were thinning and becoming more transparent, revealing slower-moving layers of gas below.

During these first few days, astronomers did not look only at the supernova itself; they also examined their records of past observations. The Large Magellanic Cloud, being such a nearby galaxy, had been studied extensively over the years. On preexplosion photographs of the region, astronomers found a 12th-magnitude star at the approximate site of SN 1987A. In 1969 Nicholas Sanduleak of Case Western Reserve University had listed it in a catalog of luminous stars as No. 202 in a strip of sky 69 degrees south of the celestial equator. Within a few days of the explosion, Sk −69 202 became a prime candidate for the star that had exploded.

This presented something of a puzzle. From the few spectra that had been taken before the blast, Sk −69 202 was a blue supergiant, about fifty times the diameter of the sun. It shouldn't have been. Most theoreticians thought that Type II supernovae resulted from the explosion of red supergiants, which are ten times bigger: as much as a thousand times the diameter of the sun. In Julian Huxley's poem, the epigram to this chapter, the reference to "stars of red and angry hue" expresses the prevailing wisdom. It was true that this wisdom was based on theory alone, and a theory is no better than the observational data that go into it. Except for one very strange and atypical supernova, 1961V, no one had ever examined a star prior to its later supernova explosion. (1961V had a light curve unlike any other supernova. There was practically no information about the star that blew up, other than that it varied erratically in brightness.) Still it was disquieting that Sk −69 202 didn't conform to expectations.

But was the Sanduleak star actually the precursor of the supernova? Kirshner, Sonneborn, and other observers on the IUE suspected for a while that it wasn't. After about a week of ultraviolet observations, the broad-lined ultraviolet spectrum of the supernova faded and vanished. Yet the IUE still detected some ultraviolet light coming from the direction of the supernova. In place of the spectrum of the supernova, Kirshner and colleagues saw the overlapping narrow-lined spectra of two faint stars, both of which looked like blue supergiants. Perhaps, they reasoned, one of these two stars was Sk −69 202, and the other was a close companion. Sure enough, a quick look at preexplosion photographs of Sk −69 202 revealed a second fainter object about three arc seconds to the northwest of the Sanduleak star. It looked like both Sk −69 202 and this second star were still going strong. The precursor of the supernova must have been a third (probably red) star, not visible on the early photographs.

Yet this interpretation didn't survive for long. In the ensuing weeks, several groups of astronomers carried out extremely precise measurements of the positions of both the supernova and Sk −69 202. The two positions agreed to within a small fraction of a second, making it very likely that supernova and star were the same object. At the same time, Nolan Walborn and Barry Lasker at the Space Telescope Science Institute in Baltimore took a closer look at the preexplosion photographs. Using elegant computer techniques, they showed that there were really three stars present, blended together by the blurring

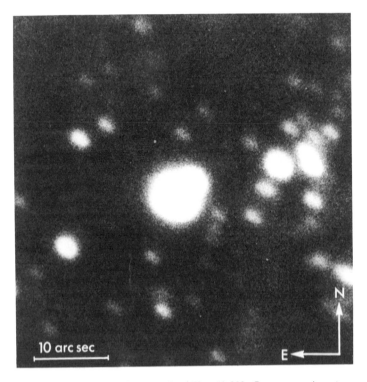

FIGURE 64: A preexplosion photograph of Sk −69 202. One companion star can be seen at the upper right edge of the blurry central image, but the second companion, at the lower left, cannot be distinguished. (European Southern Observatory.)

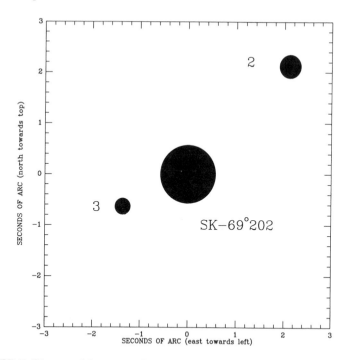

FIGURE 65: Diagram of the stars in the vicinity of Sk −69 202. Almost the entire area of this chart would lie within the blurred image of the star in the center of the previous figure. Earth's atmosphere makes it impossible to see stars as sharply as you see here. (The sizes of the star's disks have been exaggerated to illustrate their relative brightness.)

of Earth's atmosphere. The accompanying diagram shows the relative locations of the stars. Sk −69 202 is flanked by a fainter star (star 2) three arc seconds to the northwest and by another fainter star (star 3) about a second and a half to the southeast. Carefully examining the separation between the overlapping spectra from the IUE, Kirshner and Sonneborn realized that the two spectra they were seeing were the spectra of stars 2 and 3, not Sk −69 202 and star 2 as they had originally believed. Sk −69 202 was gone from the scene. It was the star that exploded.

There was no longer any way to deny that the unexpected had occurred. A blue supergiant had become a supernova. Now it remained to observe the event and to try to explain it.

MUCH ADO ABOUT ALMOST NOTHING—THE NEUTRINO BURST

More revelations followed in short order. The most spectacular, by far, was the detection of neutrinos from the initial collapse of the Sanduleak star. For decades astronomers had suspected that 99% of the energy of a Type II supernova was emitted in the form of neutrinos. But despite the enormous number of neutrinos predicted—10^{58} should be emitted in a single blast—few scientists had much hope of detecting them here on Earth.

The neutrino is one of the most insubstantial and elusive particles in nature. As far as we have been able to measure, it has no mass. If the mass of the neutrino is precisely zero (which, as we shall see, is not absolutely certain), then neutrinos travel at the speed of light, just like photons (particles of light). But neutrinos are more elusive than photons, because they do not interact with other matter very often. For example, the multitude of neutrinos generated by nuclear reactions in the solar core pass through a half million miles of sun as if it were a thin sheet of glass, emerging in a matter of seconds. They reach Earth in eight minutes and pass through it with no resistance, either. About 500 billion neutrinos from the sun fall on each square inch of ground in a second. Our bodies are pierced by them unceasingly, day and night. They leave not a trace.

Neutrinos from a terrestrial source were first detected by Fredrick Reines and Clyde Cowan at a Savannah River nuclear reactor in the 1950s. They were only able to identify the particle because the reactor produced so many neutrinos and because they were patient and careful experimenters. One in a billion billion neutrinos passing through their detector was actually recorded. Shortly thereafter, in the early 1960s, Raymond J. Davis established the first neutrino observatory, designed to detect neutrinos from the sun. Located 1.5 kilometers below ground in a Homestake, South Dakota gold mine, the Davis device was a tank containing 400,000 liters of chlorine-rich dry-cleaning fluid. The surrounding rock protected the tank from cosmic rays and other particles that might trigger the detector, but neutrinos, even those coming up through Earth from the other side, were not stopped at all. They passed through Earth, through the tank, and continued on out into space. A very small fraction of the neutrinos did not make it through, and these were the ones Davis recorded. On rare occasions, a neutrino would hit a chlorine atom in the tank, converting it into an atom of radioactive argon. Every few weeks, Davis would flush out the tank and, using a radiation counter, determine the number of radioactive argon atoms present.

Even with the enormous number of solar neutrinos passing through it, Davis only expected to count a half dozen argon atoms each day. In almost two decades of operation, he has measured about a third the expected number of neutrinos—a puzzling result that physicists are still struggling to explain. Perhaps we don't understand how the sun generates its energy as well as we think we do; or perhaps the neutrino has properties we haven't accounted for. But few scientists doubt the elegance and validity of Davis's method: he has established a science of neutrino astronomy.

A neutrino observatory like this turns our conceptions of astronomy upside down. To shield it from false alarms (cosmic rays hitting it from above), it is located underground, not on a high mountain. Rather than straining for every bit of light, it operates in perpetual darkness. Because only neutrinos can pass through the entire Earth, whereas an occasional cosmic ray may penetrate the mine from above, the neutrino detector is more effective looking down than looking up.

Davis's neutrino telescope—huge though it is—should have been incapable of detecting the neutrinos from a supernova. As a radio is tuned to receive only one channel or wavelength, his detector is tuned to receive neutrinos produced at temperatures like those deep inside the sun. The neutrinos produced in the even hotter core of a supernova should zip right through the tank without stopping, without producing enough argon to be detected. (And this was the case: Davis reported no excess argon atoms in the tank during the week of February 23, 1987.) With the detection of supernova neutrinos in mind, physicists in the early 1980s put into operation a number of "second-generation" neutrino detectors around the globe. Though more sensitive to supernova neutrinos than the Davis experiment, they were still limited to supernovae in our galaxy—or perhaps one of the nearest galaxies like the Large Magellanic Cloud. Luckily for astronomers, that's just where the 1987 supernova occurred.

In February 1987, the two largest neutrino detectors in the world were the Kamiokande II and the IMB. Kamiokande II, operated by a team of Japanese and American scientists, was located in a chamber in a lead mine beneath the city of Kamioka, Japan. The IMB, larger in physical dimensions but a bit less sensitive to neutrinos, was run by a group of physicists from over a dozen American institutions, headed by the University of California at Irvine, the University of Michigan, and the Brookhaven National Laboratory (hence IMB). The IMB group housed its detector under Lake Erie, in a large salt mine owned by the Morton Thiokol Corporation.

Both the IMB and the Kamiokande detectors were similar in design. They were large tanks of extremely pure water, sealed from outside light, and surrounded by ultrasensitive light detectors called photomultiplier tubes. The IMB detector, for instance, has a roughly cubical tank about 65 feet on a side filled with 7000 tons of water and surrounded by 2048 photomultiplier tubes. The scale of the tank may seem enormous to us, but not to the elusive particles it is trying to capture. Neutrinos entering the water usually pass right through, but one in every thousand trillion will strike an electron or a proton in the water. If the neutrino strikes an electron, the electron recoils at high

FIGURE 66: A graduate student at the University of Tokyo examining the photomultiplier tubes in the tank of the Kamiokande neutrino detector. When the detector is operating, the entire cylindrical tank, 16 meters high and 16 meters in diameter, is filled with water. The photomultipliers detect light produced by particles passing through the tank. (University of Tokyo–Kamiokande.)

speed. If the neutrino strikes a proton, a fast-moving particle called a positron is produced. In either case, the rapidly moving particles emit a cone-shaped wave of blue light, a phenomenon called Cerenkov radiation, which is picked up by the many photomultiplier tubes around the periphery of the tank and converted to electrical signals. The accompanying photograph shows a computer display of the flash of light produced by a neutrino recorded by the IMB.

Neither IMB nor Kamiokande had been designed primarily to look for neutrinos. When theoretical physicists in the late 1970s predicted that protons might be unstable, there was a great rush to put the predictions to a test. The proton had long been thought to be a fundamental, everlasting bit of matter. If protons could decay into something smaller, the atoms on which all matter was built might in time disappear: the universe might be built on shifting sands.

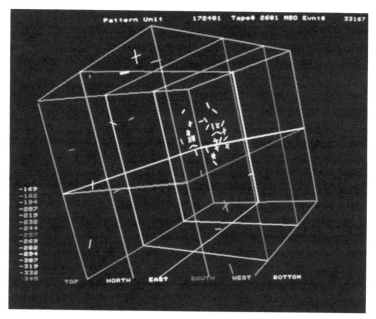

FIGURE 67: The IMB detector presents its data on a computer screen like this. The display shows a line drawing of the water tank, and x's indicate which photomultiplier tubes have detected light. The circle of x's on one wall of the tank is the signature of a neutrino from SN 1987A. (The IMB Collaboration.)

To see if this were so, large particle detectors—among them Kamiokande and IMB—were constructed to look for flashes of light from the disintegration of protons in water molecules. After several years of operation yielded no positive results, the experimenters concluded that the proton did not decay nearly as fast as predicted. So both the IMB and Kamiokande teams turned to other projects, the detection of neutrinos from supernovae among them. Some of the photomultipliers had to be made more sensitive for the project, but the modifications were relatively easy to carry out. And SN 1987A provided a spectacular payoff.

With the first news of the supernova, scientists at both Kamiokande and IMB began to look at their records. It took time to analyze the data, and it was not until two weeks after the blast that the news was announced. The neutrino pulse was there all right. Kamiokande had recorded 11 neutrinos in less than 15 seconds beginning at 7:35 a.m. Universal time (the time in Greenwich, England) on February 23, almost a day before Shelton's sighting, and about three hours before McNaught's first picture. At exactly the same time, the IMB had recorded 8 neutrinos. There was no mistaking it: even a single neutrino a day was a rare event in either detector.

Of the 10^{58} neutrinos produced by SN 1987A, about a hundred thousand trillion passed through the two detectors within a few seconds' time. Of these, only 19 were detected. But those 19 neutrinos signaled a major astronomical discovery. Taking into account the difficulty of snagging a neutrino in flight, 19 neutrinos was just the number you would expect to detect from the collapse of a Type II supernova as far away as the Large Magellanic Cloud. The neutrinos that set off sparkles in the IMB and Kamiokande tanks were produced deep in the heart of an exploding star—in effect, we were seeing the interior of an exploding star for the very first time.

There was actually a report of neutrinos from SN 1987A that predated the Kamiokande and IMB results, but it may well have been a red herring. Shortly after Shelton's discovery was announced, scientists at a neutrino observatory run by Italian and Soviet physicists in a cavern under Mont Blanc reported the detection of neutrinos from the supernova. The Mont Blanc detector consists of 90 tons of a liquid scintillator, a material that produces weak flashes of light when particles go through it. At about 3 a.m. Universal time on the night of the 23rd of February, almost 24 hours before Shelton's sighting, five flashes, apparently neutrinos, were detected in the scintillator. This was, claimed the Mont Blanc scientists, the first detection of the supernova, a vanguard of the burst of light that was to come.

The Kamiokande and IMB detections, which agreed so well with one another, cast the Mont Blanc claim in doubt. They occurred over four hours after the Mont Blanc signal. Nothing was detected by either IMB or Kamiokande during the Mont Blanc event. Moreover, according to calculations, the Mont Blanc detector should not have been sensitive enough to respond to a pulse of neutrinos from a superova as far away the Large Magellanic Cloud. The consensus is that only the Kamiokande and IMB events signal the collapse of the core of Sk −69 202 and the formation of a neutron star. The Mont Blanc pulses remain a puzzle. Though some have tried to explain them as a by-product of the collapse, most astronomers regard them as just a statistical fluctuation, a sort of "noise" in the recording system.

What did we learn from the neutrinos? Most important, we learned that the scenario of core collapse for Type II supernovae is generally correct. Just as astronomers can measure the luminosity of a star by observing how much light they receive from it, they could also measure how much energy the supernova emitted by measuring the energies of the neutrinos we received. From the 19 neutrinos detected, we calculate that about 10^{53} ergs of energy was emitted by SN 1987A in the form of neutrinos, just as predicted by the theory of core collapse. This is more energy than an entire spiral galaxy gives off in a year's time. We can even determine the temperature of the supernova core at the time the neutrinos were emitted: over ten billion degrees. Thus, the neutrinos give us a glimpse into the interior of a hellish fireball. It verges on the miraculous—19 flashes of light in subterranean darkness reflect the blazing heart of a supernova. The physicists at IMB and Kamiokande had reason to congratulate themselves.

The detection of neutrinos from SN 1987A may also have told us something fundamental about the nature of the neutrino itself: its mass. For many years physicists have been unsure whether the mass of the neutrino is precisely zero, or simply smaller than we are able to measure. Because of the difficulties of measuring such insubstantial particles, we have only been able to say with certainty that the neutrino is no more massive than about 20 electron volts (a unit we use to measure very small quantities of matter or energy), about 1/25,000 as much as an electron. That's not very much, but neutrinos are so abundant in the universe, outnumbering electrons about 100 million to 1, that even if individual neutrinos have a very small mass, the total amount of neutrino mass in the universe could be quite large, large enough to affect the future expansion of the galaxies. So the question of the precise mass of the neutrino is of considerable interest to astronomers and physicists.

Neutrinos from a supernova provide information about their mass in the following way. If neutrinos have no mass at all, then Einstein's relativity theory tells us that whatever their energy, they all move at the same speed, the speed of light. Assuming that a flood of massless neutrinos leave the supernova at the same time, they all will arrive at Earth at the same time. (Particles of light, photons, behave just this way. More energetic photons—X rays, for instance—travel at the speed of light just like less energetic photons such as radio waves.) On the other hand, if neutrinos have even a tiny bit of mass, then relativity tells us that the more energetic neutrinos should travel faster than less energetic neutrinos and should arrive here first.

The energies of incoming neutrinos can be estimated from the flashes of light they produce in the IMB and Kamiokande tanks. From these data one could, in principle, make a simple decision: If neutrinos have no mass, their time of arrival should not be related to their energy, whereas if neutrinos have a small mass, the more energetic ones should have been the first detected.

Scientists have nevertheless found it difficult to reach a consensus about the neutrino mass. Looking over the records of the neutrino detectors, it is evident that the 19 neutrinos from SN 1987A did not arrive all at once. They were spread out over a period of about 10 seconds. It is not a simple task, however, to determine from the data whether there is a relation between the energy and the time of arrival. A major difficulty is that the neutrinos released in a supernova don't all leave it at precisely the same time. Because they take some time dodging atoms on their way out of the exploding star, it takes about 10 seconds for them all to get away. Thus, most of the spread in neutrino arrival times is probably due to the supernova blast itself rather than to any difference in the speeds of the neutrinos.

That didn't stop dozens of astronomers and physicists from trying to estimate the mass of the neutrino. In the first months after the blast, scientists circulated hundreds of informal preprints about the neutrinos; speculations far outnumbered the 19 neutrinos detected. Many theorists tried to present complex statistical arguments showing a trend for more energetic neutrinos to arrive sooner than less energetic ones. But 19 neutrinos are a very small sample, and the best estimates from the data yield only approximate values. From the spread in arrival times of the supernova neutrinos, according to best estimates, we can say no more than that the mass of the neutrino is less than about 15 electron volts. It may be precisely zero. We would have had to detect more of them to make a less ambiguous determination. But we cannot repeat the experiment until the next nearby supernova goes off.

The flurry of excitement over the mass of the neutrino, even in the face of difficulties, points out how much importance astronomers give to the problem. If the neutrino mass were high, neutrinos might exert a sizable gravitational force on objects in the universe. Astronomers already suspect, from observations of the motions of stars in galaxies, that there is a lot of unseen matter in the universe that exerts gravitational force. But they are not sure what it is. It does not emit light—hence it is called "dark matter"—but beyond that its identity can only be guessed. Among the guesses: black holes, rocks the size of baseballs, and exotic subatomic particles called wimps, winos, and axions. Neutrinos are also likely explanations for the dark matter, but only if their mass is not zero.

If neutrinos are massive enough, they may produce enough gravitational force to slow and stop the expansion of the universe. Present estimates of the amount of visible matter in the universe indicate that there's not enough to slow the expansion: by this accounting the universe will expand forever. But if neutrinos have mass, the story could be quite different. Add the unseen neutrino mass to the mass of visible stars, and there might be enough matter to force the universe to stop expanding and begin contracting at some time in the distant future. If the mass of the neutrino is no higher than 15 electron volts, neutrinos probably cannot cause this to take place. But the 19 neutrinos from the latest supernova don't allow us to make any ironclad statements. SN 1987A might have told us the ultimate fate of the universe, if only we had been able to detect neutrinos more effectively.

OBSERVATIONS AND THEORIES

After the first rush of excitement, observations of SN 1987A settled into a routine. Observers in the southern hemisphere measured the light curve, watched changes in the spectrum, and tried to detect the supernova at a range of wavelengths from long-wavelength radio waves to short-wavelength X rays and gamma rays. Astronomers met at a series of international meetings, and shared their findings by circulating papers to friends and posting messages on electronic bulletin boards. Robert Kirshner, after several months of hectic activity, expressed the feeling of elation mixed with exhaustion that many astronomers shared: "The supernova is very much like having a baby—at first it's great fun and it becomes even more rewarding, but it sure is hard work!" Surprise changed to understanding as the data accumulated, and as of this writing (January 1988), we believe we have a fairly good overall picture of the causes and consequences of this particular supernova explosion.

For the first few months after its discovery, there were almost daily bulletins of the changes of light from the supernova. It rose swiftly to about magnitude 4.2 within a week. After a few days' hesitation in early March it began to rise in brightness again, and continued to rise slowly until mid-May, at which time it had reached magnitude 2.9, about as bright as the stars in the Little Dipper. (This was over a hundred times fainter than Tycho's supernova of 1572.) For several weeks it hovered at this level.

This was as bright as it was to get. Other Type II supernovae show similar plateaus in their light curves. But SN 1987A was a bit different; it was about two magnitudes (6.3 times) fainter than most other Type II supernovae at maximum. For a while the faintness of the supernova presented a problem to theorists, but now they believe they understand it. Besides generating neutrinos, the energy produced by a supernova goes partly into throwing off matter and partly into heating the matter and making it shine. Because the star that exploded, a blue supergiant, was smaller than most Type II precursors, the explosion had to lift gases farther to get them away from the surface of the star. Thus, more of the energy of the blast was used to tear the star apart, and less energy was available to provide visible light, making the supernova fainter than average.

By mid-June the light from the supernova was falling again, and by the end of June it had settled into a slow but steady decline. Remarkably, the decline was exponential in nature, just like the decay of radioactivity from an unstable nucleus. An exponential decline in light had been seen in other supernovae (mostly Type I), and had, in fact, been linked with radioactivity.

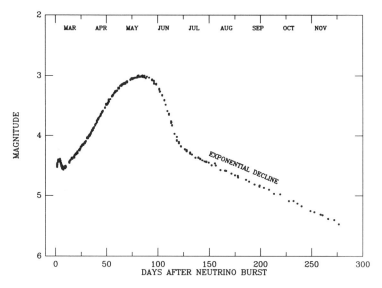

FIGURE 68: Light curve of SN 1987A. The magnitudes were measured with a light sensor aboard the International Ultraviolet Explorer satellite. Time is measured in days from the time neutrinos were detected by Kamiokande and IMB. (Data provided by Robert Kirshner and Eric Schlegel.)

The exponential fading of the supernova may be indirect evidence of elements in the making. Stanford Woosley of the University of California at Santa Cruz and David Arnett of the University of Chicago have independently generated computer models of the blast that reproduce the light curve of SN 1987A rather well. They believe that SN 1987A developed from a star about 15 times the mass of the sun with a small, dense core of iron. It became unstable and exploded, forming a sizable amount of radioactive nickel-56 (7% of the total mass of the sun), and blowing it out with the rest of the debris. The decay of nickel into cobalt-56 and then into stable iron-56 produced gamma rays (energetic photons) that heated the surrounding gas to incandescence. After mid-June, this radioactive heating was the main source of energy that made the supernova shine. Gradually the radioactivity declined, and with it the light from the supernova. The exponential decline in the light of SN 1987A between June 1987 and the end of the year precisely matched the rate expected from the decay of cobalt-56, which halves in intensity in about 77 days. This was striking evidence that a supernova could generate heavy elements.

There was additional indirect evidence to support this conclusion. Gamma rays emitted by radioactive nickel and cobalt should be scattered by the debris of a supernova, producing X rays as well as light. In midsummer, not long after the exponential decline in brightness began, the Japanese X-ray satellite, Ginga, began to detect X rays from the supernova. At the same time an X-ray detector on board the Soviet space station MIR noticed an increasing signal from the supernova. Again, the gamma rays seemed to be the cause.

Both the exponential light curve and the X rays were only indirect evidence of radioactivity. Could the gamma rays be detected directly? Gamma rays from nickel-56 and cobalt-56 have a characteristic wavelength and, just like the lines in the spectrum of a gas, should be easily recognizable. In the early months after the blast, however, the gamma rays cannot make their way out through the blanket of debris that surrounds the supernova. As the debris thins, however, revealing the interior of the expanding cloud, the gamma rays should appear like the sun through a clearing fog. Theoreticians like Stan Woosley and Richard McCray of the University of Colorado predicted that gamma rays would be measured by satellite detectors sometime around the first of January 1988.

The supernova, however, surprised them pleasantly by showing its gamma rays about 4 months ahead of schedule. Just before Christmas 1987, a group of scientists announced the analysis of data from a gamma-ray detector aboard the Solar Maximum Mission, a satellite launched in 1980. Since late August, they reported, the satellite had recorded gamma rays from the direction of the supernova, at precisely the wavelengths expected from the decay of cobalt-56. The unusual strength of the signal at this early date might be explained if the shell of expanding material from the supernova had broken into tendrils and fragments. This would expose the interior to view even before the outer edges of the cloud had thinned and become transparent.

Whatever the reason for their early appearance, this first-ever detection of gamma rays from the radioactive by-products of a supernova provided us with striking evidence that a supernova could forge the elements. It was a fitting Christmas gift to the astronomical community, and a remarkable ending to a year of excitement.

CLUES IN THE SPECTRA

The spectrum of the supernova has provided additional insights into the process of element formation in massive stars. In the early spectra of the supernova, hydrogen lines were the strongest features by far. But as the debris thinned and became more transparent, the lines of other chemical elements began to appear. Among them were lines of iron, calcium, and strontium, similar to heavy elements seen in other Type II supernovae. Some elements were overabundant compared to the average composition of material in the universe. Barium, for instance, which is produced in the s process during the late life of a massive star, appeared stronger than normal. Here, we suspect, we are seeing recently created atoms being scattered into interstellar space by the explosion. The supernova acts as a distributor as well as a creator of the elements.

Robert Kirshner and his collaborators see evidence that even before it became a supernova, Sk −69 202 was shedding heavy elements into space. Early in fall 1987, narrow spectral lines began to appear in the ultraviolet spectrum of the supernova. The emission lines were too narrow to come from the turbulent gases ejected by the blast. Instead, suggested Kirshner, they represent a shell of material about a light-year in diameter surrounding the star. The material appeared to be particularly rich in nitrogen, like the material around the Eta Carinae Nebula and the slow-moving material surrounding the Cassiopeia A supernova remnant. Gas like this, enhanced in nitrogen by the CNO energy-producing process (which takes place in massive stars like Sk −69 202), should be expelled by a star in its years as a red supergiant. Perhaps Sk −69 202 was a red supergiant several thousand years ago. Perhaps it blew off this nitrogen-rich material and became a blue supergiant, after which it exploded. This material produced the narrow lines Kirshner detected.

The spectrum of the surrounding material didn't become evident the instant the star exploded. As Kirshner sees it, the initial explosion of Sk −69 202 produced a flood of ultraviolet radiation that traveled outward at the speed of light until it encountered the shell of previously expelled material. The ultraviolet light caused the gases in the shell to fluoresce, just like ultraviolet light can make a "black light" poster shine with eerie colors. Because it took time for the ultraviolet radiation to reach the shell of material, we had to wait about six months before the nitrogen-rich spectrum appeared. Now, illuminated by the star's last burst of brilliance, we see the material that Sk −69 202 expelled many years ago. It all fits together neatly, and matches expectations gathered from the fragments of other explosions like Cassiopeia A.

THE MYSTERY SPOT

In many ways, SN 1987A has behaved just about as we expected. But there are some abiding mysteries that are giving astronomers sleepless nights. One of the most baffling is the nature of a bright dot of light that astronomers detected about a twentieth of a second of arc from the supernova, just a month after it first exploded. The dot was a hundred times brighter than anything that had been there before the blast, and it was presumably created by the supernova. Yet there is no convincing way to explain its sudden appearance. Astronomers have termed it "the mystery spot."

The mystery spot, because it is so close to the supernova itself, could not have been detected by ordinary photographic techniques. Light from the supernova, like all starlight, is smeared out by the turbulent motion of Earth's atmosphere, and all we see is a smudge of light about a second of arc in diameter. Nearby stars are blended into the smudge, and can't be distinguished as separate objects. In recent years, however, a number of enterprising astronomers have employed a new technique called speckle interferometry to eliminate the blurring caused by atmospheric motions. Instead of taking a single long exposure of a star, they take a series of very short exposure (1/1000 of a second) pictures and combine them in a computer. Each picture is taken so quickly that the motion of Earth's atmosphere doesn't have time to blur the image. The amount of light collected in each picture is very small, but the computer can add thousands of such frames together, with the effects of Earth's atmosphere virtually eliminated, to produce a sharp, well-defined picture.

At the Harvard–Smithsonian Center for Astrophysics in Cambridge, Massachusetts, Costas Papaliolios, Peter Nisenson, Margarita Karovska, and Robert Noyes had been working on an advanced speckle system for making high-definition images of the sun. When the supernova went off, they realized that their instrument, which could see finer detail than any ordinary camera, was ideal for watching the early development of a supernova remnant. They quickly

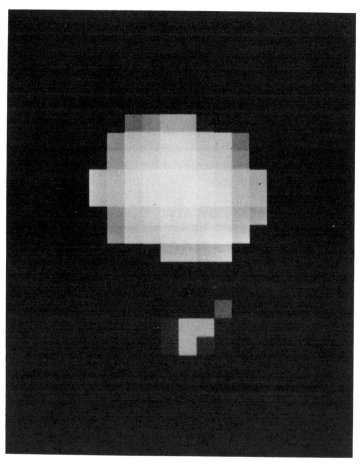

FIGURE 69: Image of the "mystery spot" near SN 1987A obtained by speckle interferometry. Individual picture elements (pixels) assembled by the computer processing
appear as small squares. The supernova is the large blob of pixels near the center, and
the "mystery spot" below it is the smaller group of about a half-dozen pixels. The
separation between the supernova and the spot is about 1/20 of an arc second.
(Courtesy of Peter Nisenson, Harvard–Smithsonian Center for Astrophysics.)

arranged to attach their equipment to a large telescope at Cerro-Tololo Observatory in Chile. By getting ultra-high-resolution images, they hoped to measure directly the size of the supernova's expanding shell and, incidentally, determine whether the Sanduleak star had indeed survived. (At this time it was still unclear which star had exploded.) On March 25, and again on April 2, they carried out observations using a sensitive camera that Papaliolios had designed several years earlier.

To produce an image of the supernova and its close neighbors, the Cambridge astronomers combined 60,000 very short exposures on a computer. The accompanying picture shows the surprising result. The supernova itself is the large, bright object at the center. About a twentieth of a second to the southwest is another object—the mystery spot. No one has yet been able to explain it.

The mystery spot was not there before Sk −69 202 exploded. That much is clear. If it seems faint on the computer image, that is only in comparison to the adjacent supernova. Looking at preexplosion photos of the region, we can see that the mystery spot is over six hundred times brighter than any star in the vicinity before the blast, including Sk −69 202 itself. (Do not confuse the mystery spot with either of the two nearby stars we discussed earlier. It is much closer to Sk −69 202 than either star 2 or star 3 in the preexplosion images.) Yet it is hard to understand how the supernova itself could have produced such an object. The observed separation of a twentieth of a second of arc means that the mystery spot was about 500 billion miles away from the supernova when it was detected. If it was a blob of stuff ejected by the blast, it would have had to move at exceptional speeds, about a third the speed of light, to move this far in a month's time. It is difficult to imagine a supernova expelling material at this high a speed.

Even more confounding is its brightness. Though the mystery spot was only 10% as bright as the supernova, that 10% represents an enormous amount of energy. Suppose that the spot was simply a patch of nearby gas and dust lit up by the supernova. Even if it reflected all the light that hit it, like a highly polished mirror, it would have to be enormous to reflect as much light as we receive—big enough to show up as more than a dot on the speckle images. To get around this difficulty, some have suggested that the spot was produced by a concentrated beam of energy radiated by the supernova directly at a nearby interstellar cloud, like a searchlight on the cloud deck over a city. It could then be a somewhat smaller object. Others suggest that the spot was a second supernova triggered in a nearby binary companion star, just as the falling of one domino can cause an adjacent one to topple. But no one seems certain how an external event like a nearby supernova can trigger a star to explode on short notice. All the suggestions are highly contrived. The truth is, the mystery spot is simply an enigma.

To deepen the mystery, the spot has since disappeared. It was present, apparently, in mid-April 1987, when an independent group of observers from Imperial College in London obtained speckle measurements of the supernova. But when Papaliolios, Nisenson, and Karovska reobserved the supernova in May and July, there was no trace of the object. It may have faded below the limits of detection of their instrument (about 40 times fainter than the supernova). Without additional observations to go on, the mystery spot may remain an unsolved problem for some time to come.

WHAT DO WE KNOW SO FAR?

Aside from the mystery spot, observations of SN 1987A have been able to provide us with a remarkably satisfying picture of what went on in the Large Magellanic Cloud. From the observations we can piece together the following general story: Sk −69 202 began its life as a main sequence star about 20 times as massive as the sun, burning hydrogen deep in its interior. After about 10 million years, it began to burn out the hydrogen in its core, and it expanded to red-giant size. Soon it was burning heavier and heavier elements. At the same time, it was losing several solar masses of its surface material into space, surrounding itself with a cloud of heavy-element enriched gas. By the end of its life the star was down to perhaps 15 solar masses overall and looked, from the outside, like a blue supergiant. But deep inside it had developed a core of iron a little more massive than the sun.

Eventually, the iron core of Sk −69 202 reached a critical mass of about 1.4 times that of the sun. Unable to support its own weight any longer, the core collapsed, forming a neutron star, and emitting a powerful burst of neutrinos that reached Earth at 7:35 a.m. Universal time, February 23, 1987. A day later, as the blast broke through the surface of the star, the supernova began to emit noticeable amounts of light. The outer layers of the star were ejected, including a large amount of radioactive nickel-56. The radioactive nickel later decayed to cobalt and iron, kept the expanding debris from the supernova hot and shining, and produced X rays and gamma rays as well. The ejected gas is thinning now, and will in time become a diffuse, wispy supernova remnant.

The neutron star left behind by the blast may very well be spinning very rapidly and may have a very strong magnetic field. If so, as the debris clears, we may see a pulsar blinking rapidly at the center of the remnant. We haven't seen it yet, but if we do it would be a remarkable climax to the study of SN 1987A. Observation of a pulsar would represent a striking confirmation of the theory of Type II supernovae—stellar death and transfiguration observed from start to finish.

WHAT'S NEXT?

Heaven has smiled on astronomers. If supernovae occur two or three times a century in a large spiral galaxy, they should occur every thousand years or so in a smaller galaxy like the Large Magellanic Cloud. What a stroke of good fortune that SN 1987A occurred so close by and so soon after Baade and Zwicky's farsighted paper on supernovae!

The good fortune may not be over yet. After a year, Supernova 1987A remains visible to the naked eye, a very bright object for most astronomical instruments. If it continues to decline at the present rate, astronomers may be able to examine its optical spectrum for five years or more, giving us a continuous record of the early life of a supernova remnant. Satellite observations and radio observations will continue to add useful information for decades or even centuries. In addition, Papaliolios, Nisenson, and their collaborators continue to make speckle observations of the ejected cloud around the supernova. They have not seen the mystery spot again, but they already see the supernova growing in size as it expands into space, and they expect to reobserve it at regular intervals.

What about the neutron star the supernova left behind? It is presently hidden within the cloud of debris, but as the cloud thins, it should become visible, perhaps as a small dot of X-ray emission. If the neutron star is a pulsar, there is about a 30% chance its lighthouse beam of radiation will cross our line of sight. Perhaps in a year or several years (no one is certain), we will see flashes of light or radio waves. Astronomers are already watching, but they see no sign of pulses so far.

A pulsar may also reveal its presence indirectly, by augmenting the visible light we see from the debris around the supernova. The radiation of a pulsar can heat the expanding gases from the blast, causing them to shine. We have noted that most of the heating of these gases seems presently to come from the decay of radioactive cobalt, but as the radioactivity declines, heating from the pulsar may account for a larger fraction of the light. If there is a pulsar to provide energy, we may expect to see the fading of the supernova gradually slow or stop. The pulsar will keep it glowing. Otherwise the decline in light should continue until the remnant fades from sight.

Bradley Schaefer of Goddard Space Flight Center predicts that we may see a light echo from SN 1987A just as we saw a light echo around Nova Persei in 1901. In several years' time, the light from the initial flare-up of the star will be reflected from clouds of interstellar gas and dust several light-years away from the star. Like the echo from Nova Persei, the echo from this supernova would appear as a growing arc of light surrounding the site of the vanished Sk −69 202. If conditions are right, says Schaefer, the echo should even be visible for decades, using nothing more elaborate than a pair of binoculars.

Schaefer's predictions may well be borne out. Early in 1988 astronomers began to notice two faint arcs of light centered on the site of SN 1987A, which had not appeared in earlier photographs. When first seen, the diameters of the two arcs were about 1% the diameter of the full moon, but over a month and a half's time they were observed to be steadily growing in size, promising a fine showing in the years to come.

Even later, suggests Kenneth Brecher of Boston University, we may see a second burst of light from SN 1987A. The cloud of debris expelled by the blast, moving at a tenth the speed of light, should eventually collide with the slower-moving shell of gas expelled when Sk −69 202 was a red giant star. (This is the nitrogen-rich shell detected by Kirshner as a set of narrow spectral lines.) The collision will heat the gas, causing it to glow. From Earth, it will look like the supernova has lit up once again. Though it will only reach a ten-

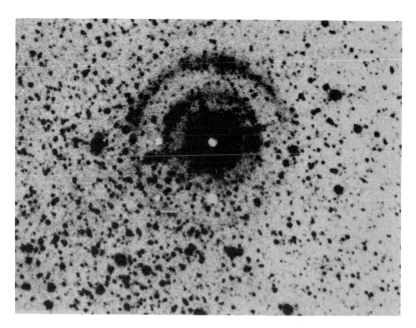

FIGURE 70: Light echoes from SN 1987A. Two concentric rings of light surrounding the supernova were first recognized by Dr. Arlin Crotts, a postdoctoral fellow at the University of Texas, on March 3, 1988 using a telescope at Las Campanas, Chile. On closer inspection, astronomers found the echoes in earlier photographs. The negative image above was taken by Dr. Michael Rosa at La Silla, Chile, on February 13, 1988. Light from the supernova itself has been dimmed by an obscuring disk placed in the telescope; the "cross" extending from the supernova is an artifact of the telescope optics. Surrounding this are two concentric rings, 30 and 50 arcseconds in diameter, which are formed by light from the supernova reflecting off of two clouds of interstellar dust about 400 and 1000 light-years in front of the supernova. (European Southern Observatory.)

thousandth the brightness it had at its peak in May 1987, this reignition should be easily visible to astronomers.

Brecher notes that such "reborn" supernovae have been noted in the past. The Chinese annals report the appearance of a "guest" star in 1016 at the same position as the supernova of 1006. And Simon Marius, a student of Tycho Brahe, recorded a rebirth of Tycho's 1572 supernova in 1613. These second flashes of supernovae may be common occurrences, but unless the supernova is very close, they cannot be seen without a telescope. It was therefore fortunate for Marius that the supernova he observed was in our Milky Way, and fortunate that he was one of the first astronomers to use a telescope. Now, with the benefit of almost four centuries of technology, we are prepared to detect the aftershocks of a supernova ten thousand times more distant. If these predictions are correct, supernova 1987A will be in the news for many years to come.

CHAPTER 12

Into the Future

> And the end of all our exploring
> Will be to arrive where we started
> And know the place for the first time.
> —T. S. Eliot, *Four Quartets*

A NEW SCIENCE

Supernova 1987A is history now, joining the distinguished company of Tycho's star and the great supernova of 1054. As a result of all the attention it has received, the complexion of supernova research has changed dramatically. Until 1987, astronomers studying supernovae had a mixed bag of data to use in their work. There were ancient documents written by observers who did not recognize what they were seeing at the time. There were scattered observations of supernovae in distant galaxies, seldom discovered until days or weeks after they went off, and often too faint to study in great detail. There were studies of pulsars and supernova remnants: objects formed at dates and under circumstances that, in most cases, could only be guessed at. The story of supernovae was, for the most part, a patchwork of circumstantial evidence.

Paleontologists often reconstruct extinct animals in similar fashion—a thighbone from one dig, a skull from another, a few vertebrae from a third. They wire together a plausible-looking skeleton, and then, using a knowledge of anatomy, try to imagine what the living animal must have been like. Seldom can they check their work against a whole animal, but there are exceptions. The appearance of SN 1987A was like the discovery of an intact mammoth, flesh and all, frozen in an Arctic bog, after decades of seeing only scattered bones.

Here was the entire beast; here was a chance to see how close to reality the guesswork had brought us.

In many ways, SN 1987A confirmed what we already suspected. The neutrino burst, the eventual detection of gamma rays from cobalt-56, the chemical composition and velocity of ejected gases, all conformed to expectations. There were also surprises. Few would have bet that a blue, rather than a red star would explode. Few anticipated the relative faintness of the supernova at maximum. No one predicted the mystery spot. In the long run, many details remain to be explained.

In the balance, most astronomers feel satisfied that nothing we have seen so far will radically alter our fundamental view of what happened when Sk −69 202 exploded. The core of a massive star collapsed; a neutron star was formed; in time we shall detect a gaseous remnant. Before SN 1987A, we merely suspected that such things happened. Now we know.

Yet we must be cautious in generalizing from the experience of the supernova in the Large Magellanic Cloud. Just as a single frozen mammoth cannot tell us all there is to know about the evolution of mammals, so SN 1987A cannot tell us all there is to know about supernovae. It was a Type II supernova after all, and Type II supernovae are only part of the supernova story. Type I supernovae, which may be as common, are produced not by core collapse, but by the nuclear detonation of a carbon-rich white dwarf in a binary system (see Chapter 6). Until we can observe a Type I supernova as closely as we have observed SN 1987A, the story of supernovae will be necessarily incomplete.

Nor is it clear how much SN 1987A tells us about Type II supernovae. It seems to be one of a kind in many respects. No other Type II on record had a light curve of quite the same shape: the slow rise to maximum, the long delay before exponential decline. Few Type II supernovae seem to have been as faint. It is possible that supernovae like SN 1987A are in fact common in the universe, but go undetected because they are too faint to be seen over large distances. Or perhaps there are already one or two supernovae on record that are like SN 1987A, but have not been recognized as such. In these cases we may have missed the first few weeks after core collapse because the supernova had not yet brightened enough to stand out from the surrounding galaxy. The way to recognize more supernovae like SN 1987A may be to increase the sensitivity of our searches so that we can study more supernovae as early in their development as possible.

If we are lucky, we will catch a few supernovae in galaxies close enough to tell what kind of star exploded. It is likely that SN 1987A changed from a blue to a red supergiant over the course of several thousand years as a result of shedding its outer layers of gas. Though we know such mass loss is common among aging stars, it's not clear whether most supernovae occur among stars that have shed material, or whether red supergiants themselves can explode before much mass has been lost. It may be that the amount of mass a star loses before collapsing is a major determinant of what the explosion looks like. There may be a continuous range of such supernovae, from those that have lost little mass (ordinary Type II?), to those that have lost some mass (like SN 1987A?), to those that have lost much (Type Ib?). Again, the only way to decide this will be to observe more supernovae, and their precursor stars, in more detail.

Supernova 1987A thus not merely answers some decades-old questions about supernovae, but gives us more reasons to keep on looking. Now there are new questions to answer, and old ones that remain unresolved. We do not know when the solutions will be forthcoming, only that more observations are needed. But we can take heart. At this very moment the bursts of light from countless supernovae are speeding toward Earth. As each new supernova appears in our skies, we will be better equipped to understand what is happening.

NEW WAYS OF LOOKING

Modern technology has brought substantial improvements in our understanding of supernovae. Nevertheless, it is notable how much of the research on SN 1987A was done with relatively old equipment. Ian Shelton's discovery was made on a telescope built over a half century ago. Ultraviolet monitoring of the supernova was carried out by the International Ultraviolet Explorer, a satellite launched in 1978—ancient by the standards of a fast-moving technology. The IUE contains a rather small telescope, less than 20 inches in diameter, and it is showing severe signs of age at present. Similarly, the first detection of cobalt-56 gamma rays was made by the Solar

Maximum Mission, a satellite launched in 1980 and later repaired in orbit by the crew of the Space Shuttle.

Shelton's discovery, it must be admitted, demanded little in the way of technology. Oscar Duhalde, after all, made his first sighting with the oldest of all instruments: the naked eye. There's no need for high tech when it comes to spotting bright, nearby supernovae at optical wavelengths. Of all the astronomers at work in Chile on the night of February 23, 1987, only the most modestly equipped discovered a supernova. The others, more than likely, were watching the TV monitors of their high-tech instruments, unaware of anything outside the restricted patch of sky visible on their screens.

Observations at wavelengths outside the visible, however, require high-tech telescopes placed in space. Our reliance on aging instruments is in part due to the current crippled state of the U.S. space astronomy program. The IUE, productive but ailing, is the sole functioning ultraviolet telescope in orbit. The Hubble Space Telescope, which will provide unparalleled views at ultraviolet and visible wavelengths, awaits the resumption of Shuttle flights before it can be launched. Other satellites, like the Advanced X-Ray Astronomy Facility (AXAF), which would have given us more sensitive X-ray measurements than either the Soviet MIR or the Japanese Ginga experiments, have suffered both from delays and from cutbacks in financial support.

Within the next decade and a half, as new satellites are developed and launched, we will no doubt see a major improvement in this situation. New instruments will enable us to detect fainter objects, see them more sharply, and study them at wavelengths ranging from the infrared to gamma rays. Some of the new satellites will devote time to studying the dozen or so faint, distant supernovae that are discovered each year. (This rate may increase if the automated supernova searches are successful.) They will also look closely at supernova remnants in the Milky Way and in nearby galaxies. And when the next bright supernova appears, they will be able to detect its X rays and gamma rays earlier, follow its ultraviolet spectrum longer, measure the size of its expanding shell with better precision than is currently possible.

There will be improvements on ground-based equipment as well. In a few years, giant optical telescopes like the 10-meter Keck telescope being built by the California Institute of Technology, will increase our capability to study distant supernovae. Large arrays of linked radio telescopes spanning almost half the globe will make it possible to study the expansion of supernovae and the structure of remnants in our own galaxy and others.

For the study of bright nearby supernovae, I suspect that the most significant advance will be the development of larger and more sensitive neutrino detectors. Masa-Toshi Koshiba, a leader of the team of scientists that runs the Kamiokande neutrino detector, has proposed scaling up the current swimming-pool-sized water tanks to the size of football fields and increasing their ability to respond to bursts of neutrinos from supernovae. He envisions a worldwide network of detectors like Kamiokande and IMB, linked together by communications lines, which could serve as an early warning detector for supernovae. Such a network would alert us to supernovae even before their light reaches Earth.

Koshiba's neutrino alarm might work like this: A star collapses, producing a flood of neutrinos that escape long before the rebounding shock has a chance to reach the surface of the star. The neutrinos travel to Earth, but strike detectors on the near side of Earth a fraction of a second before they strike ones on the other side. The timing information from a half-dozen neutrino telescopes is used to determine the direction of the supernova, and astronomers begin to monitor that part of the sky for the first appearance of light from the exploding star.

Koshiba's plan is surely feasible. We successfully detected neutrinos at least three hours before we detected the light from SN 1987A. The IMB and Kamiokande groups, to be sure, only examined their data after the supernova was seen optically. (The Mont Blanc detector *did* automatically print out a notice of a neutrino pulse, but most astronomers don't believe the pulse was produced by the supernova.) Now that we know what to look for, we should be able to use the neutrinos as a first alert.

In time we may be able to detect gravitational waves as well as neutrinos from supernovae. Gravitational waves are an exotic phenomenon predicted by Einstein's theory of general relativity. When the core of a star collapses rapidly, it can produce a gravitational disturbance that travels outward from the star with the speed of light. The disturbance affects objects in its path by causing them to distort, like the distortion of a rubber ball when you squeeze it. But the distortion is very small: a gravitational wave from a nearby supernova passing through Earth causes the ground to shift less than a ten-millionth of an inch. Nevertheless, it is possible to measure these tiny tremors.

Joseph Weber, a physicist at the University of Maryland, has been a pioneer in the field of gravitational wave detection. His detectors are large metal cylinders equipped with sensors that can detect the minuscule distortions in the cylinder that occur when a gravitational wave passes by. At a meeting of astronomers in the fall of 1987, Weber announced that he had detected a possible gravitational wave signal from SN 1987A. However, most scientists believe that Weber's present detectors are far too insensitive to detect a supernova, and they view his claim with skepticism.

The prospects for gravitational wave detection may be better in the next decade. A number of other gravitational wave detectors are under construction around the world. Some use Weber-type cylinders. Others use lasers to measure the tiny change in distance between two mirrors caused by the passage of gravitational waves. When they are in operation, we may be able to sense the collapse of a distant star by the subtle twitchings it produces on our planet, light-years away from the scene of disaster.

PROSPECTS

Even with improved technology, however, we stand to learn the most from bright nearby supernovae like SN 1987A. Supernovae probably occur once every thousand years or so in the Large Magellanic Cloud, so there is not much chance of seeing another one there in the near future. The Great Nebula in Andromeda has many more stars, and should produce a supernova once or twice a century. But it is ten times farther than the Large Magellanic Cloud; the Kamiokande detector, at present, would be hard put to detect a signal from a supernova that far away. Unfortunately, the Andromeda Nebula is the nearest large spiral galaxy outside our own.

The next great supernova will be located in the Milky Way, probably within a few thousand light-years of the sun. We are long overdue. Supernovae in a galaxy the size of the Milky Way should occur slightly more frequently than once a century, yet the most recent we know of, the one that gave rise to the Cassiopeia A remnant, went off over 300 years ago. Where and when will the next one occur?

We could make an educated guess. At the heart of the Eta

Carinae Nebula, about 9000 light-years from us, lies a star perhaps a hundred times as massive as the sun. Its behavior over the past few centuries has been strikingly erratic. Edmond Halley, on a voyage to the island of St. Helena in 1677, recorded the star's magnitude as 4, several times fainter than the stars in the Little Dipper. In 1829 the star had brightened to magnitude 1, comparable to the stars in the belt of Orion. And in the spring of 1843, John Herschel (William Herschel's son), observing from the Cape of Good Hope, noted with amazement that the star in Eta Carinae had risen to magnitude −1, second in apparent brightness only to Sirius. In the 1850s, it plunged by a factor of 10,000 to about magnitude 8, well below the limit of the naked eye; since 1940, the star has been steadily brightening again.

Astronomers are not certain what is going on. But it is possible that Eta Carinae's variability is a sign of impending disaster. We know that the surrounding nebula is enriched in nitrogen, a sign that the central star has already shed some of its nuclear-processed material into space, and that it may be reaching the end of its stable life. Nevertheless, we can say no more than that Eta Carinae may become a supernova sometime within the next few thousand years. If it does, it will be brighter than the planet Venus.

Another candidate is the star Betelgeuse, a bright reddish star only 500 light-years away in the shoulder of the constellation of Orion. Betelgeuse is a red supergiant, large enough to engulf the orbit of the planet Mars if it occupied the place of our sun. If Betelgeuse became a supernova, it would be brighter at maximum than the full moon, and would remain visible to the naked eye for over a decade. As far as we can tell, this must happen sooner or later, but again, it could be many thousands of years in the future.

We really cannot tell which star will be the next to go. As likely as not, it will be one we have not even considered, or perhaps one we do not even know of. Suppose that the next supernova in the Milky Way occurs behind a cloud of obscuring material like that which hid Cassiopeia A from view three centuries ago. The neutrino detectors will record a strong pulse, the Weber cylinders will flinch, the radio telescopes will detect a signal—but only a faint flicker of light will be detected by optical telescopes. As always, astronomers will make the best of it, learning all they can from what little they see.

A hundred years may pass before there's another supernova to match the one that gave birth to the Crab Nebula or the one that made young Tycho doubt his senses. Yet I am hopeful that one clear day my phone will ring and an excited voice will say something like: "Have you heard about the bright supernova in Orion? The neutrinos came in an hour and a half before the first optical sighting. It's at magnitude −4 already. And on the rise." Light-headed and smiling happily, I'll rush off to tell the news to anyone who'll listen. Later, in the deepening twilight, we all will see a new and brighter evening star.

Bibliographical Notes

A NOTE ON SOURCES

Twenty years ago, when I first wrote about supernovae for a graduate-school seminar, a complete bibliography on supernovae fit comfortably on a pack of index cards. The bibliography on SN 1987A alone exceeds that size already, and a complete bibliography on supernovae, if it could be compiled, would be far too large to include here. In writing this book I have relied on numerous articles, in both professional journals and popular magazines, as well as interviews and informal discussions with researchers in the field. Several books and articles have been especially helpful, and I would like to note them here.

A major source on the history of astronomy prior to the 20th century was Anton Pannekoek, *A History of Astronomy* (London: Allen & Unwin, 1961). Much of the material on supernovae in ancient times was drawn from David H. Clark and F. Richard Stephenson, *The Historical Supernovae* (Elmsford, N.Y.: Pergamon Press, 1977), and the background material on Chinese science from Joseph Needham and Wang Ling, *Science and Civilization in China*, Volume 3 (London: Cambridge University Press, 1959). For material on the beginnings of modern supernova research there are many review articles and reminiscences, but I began with Fritz Zwicky's own article in Cristiano Batalli Cosmovici (Ed.), *Supernovae and Supernova Remnants* (Dordrecht: Reidel, 1974).

The technical material in the later chapters was drawn from literally hundreds of articles and monographs. Among the most readable and informative are two lengthy review articles: Virginia Trimble, "Supernovae: Part I: The Events," *Reviews of Modern Physics* 54 (October 1982), pp. 1183–1224; and Virginia Trimble, "Supernovae: Part II: The Aftermath," *Reviews of Modern Physics* 55 (April 1983), pp. 511–563. Finally, a well-balanced, semitechnical review of SN 1987A is found in David Helfand, "Bang: The Supernova of 1987," *Physics Today* 40 (August 1987), pp. 25–32.

SUGGESTED READING

Books

Isaac Asimov, *The Exploding Suns* (New York: Dutton, 1985).

David H. Clark, *Superstars* (New York: McGraw–Hill, 1984).

David H. Clark, *The Quest for SS433* (New York: Viking Penguin, 1985).

Russell Genet, Donald Hayes, Donald Hall, and David Genet, *Supernova 1987A: Astronomy's Explosive Enigma* (Mesa, Arizona: Fairborn Press, 1988).

George Greenstein, *Frozen Star* (New York: Freundlich Books, 1983).

*Minas C. Kafatos and Richard B. C. Henry (Eds.), *The Crab Nebula and Related Supernova Remnants* (London: Cambridge University Press, 1985).

*Minas C. Kafatos and A. G. Michalitsianos (Eds.), *Supernova 1987A in the Large Magellanic Cloud* (London: Cambridge University Press, 1988).

Paolo Maffei, *Monsters in the Sky* (Cambridge, Mass.: MIT Press, 1977).

Richard N. Manchester and Joseph H. Taylor, *Pulsars* (San Francisco: Freeman, 1977).

Simon Mitton, *The Crab Nebula* (New York: Scribner, 1978).

Paul Murdin and Leslie Murdin, *Supernovae* (London: Cambridge University Press, 1985).

*Martin Rees and Ray J. Stoneham (Eds.), *Supernovae: A Survey of Current Research* (Dordrecht: Reidel, 1982).

*F. G. Smith, *Pulsars* (London: Cambridge University Press, 1977).

Walter Sullivan, *Black Holes* (Garden City, N.Y.: Anchor Press/Doubleday, 1979).

Gerrit Verschuur, *The Invisible Universe Revealed* (Berlin: Springer-Verlag, 1987).

Magazine and Journal Articles

Hans A. Bethe and Gerald Brown, "How a Supernova Explodes," *Scientific American* (May 1985), p. 60.

Chaman Bhat, Chris Mayer, and Arnold Wolfendale, "In Search of the Source of Cosmic Rays," *New Scientist* (6 February 1986), p. 48.

Adam Burrows, "The Birth of Neutron Stars and Black Holes," *Physics Today* (September 1987), p. 28.

Neil Comins and Laurence A. Marschall, "How Do Spiral Galaxies Spiral?" *Astronomy* (December 1987), p. 6.

Paul Gorenstein and Wallace Tucker, "Supernova Remnants," *Scientific American* (July 1971), p. 74.

*A more technical book.

Nigel Henbest, "Supernova: The Cosmic Bonfire," *New Scientist* (5 November 1987), p. 52.

William Herbst and George E. Assousa, "Supernovas and Star Formation," *Scientific American* (August 1979), p. 138.

Ronald N. Kahn, "Desperately Seeking Supernovae," *Sky and Telescope* (June 1987), p. 594.

Francis Reddy, "Supernovae: Still a Challenge," *Sky and Telescope* (December 1983), p. 485.

Ronald A. Schorn, "Supernova 1987A after 200 Days," *Sky and Telescope* (November 1987), p. 477.

Ronald A. Schorn, "Happy Birthday, Supernova!" *Sky and Telescope* (February 1988), p. 134.

Frederick D. Seward, "Neutron Stars in Supernova Remnants," *Sky and Telescope* (January 1986), p. 6.

Frederick D. Seward, Paul Gorenstein, and Wallace H. Tucker, "Young Supernova Remnants," *Scientific American* (June 1976), p. 100.

F. Richard Stephenson and David Clark, "Historical Supernovas," *Scientific American* (December 1974), p. 100.

Richard Talcott, "Insight into Star Death," *Astronomy* (February 1988), p. 6.

Kurt Weiler, "A New Look at Supernova Remnants," *Sky and Telescope* (November 1979), p. 414.

J. Craig Wheeler and Robert P. Harkness, "Helium-Rich Supernovae," *Scientific American* (November 1987), p. 50.

J. Craig Wheeler and Ken'ichi Nomoto, "How Stars Explode," *American Scientist* (May/June 1985), p. 240.

S. E. Woosley and M. M. Phillips, "Supernova 1987A!," *Science* (6 May 1988), p. 750.

An extensive bibliography of nontechnical material on SN 1987A (and supernovae in general) is available from the Astronomical Society of the Pacific, 1290 24th Avenue, San Francisco, Calif. 94122.

Glossary

Absorption line—A dark line in a spectrum. If the spectrum is represented as a graph of intensity versus wavelength, an absorption line appears as a dip. It is produced as radiation from a hot source passes through a cooler gas, which absorbs at specific wavelengths.

Accretion disk—A flat disk of infalling material whirling around a black hole or a neutron star.

Allende meteorite—A meteorite, of the rare old type known as a carbonaceous chondrite, that fell in Mexico in 1969.

Andromeda Nebula—A nearby spiral galaxy, also known as M31.

Angstrom—Unit of length used to measure wavelengths of light. It is defined as 10^{-8} (0.00000001) centimeter.

Arc second—An angle equal to 1/3600 of a degree; roughly, the apparent size of a dime at a distance of a mile.

Asterism—A group of several stars that appear very close together in the sky.

Beta decay—Type of radioactive decay in which a nucleus emits an electron and an antineutrino, thus transforming one of the neutrons in its core to a proton.

Binary star—A pair of stars in orbit around one another.

Black hole—An object so small and dense that no light can escape from it.

Blue supergiant—A star whose surface temperature is very hot (tens of thousands of degrees) and whose diameter is about 50 times that of the sun. A late stage in the life of a massive star.

Brightness—The rate at which energy is received from a source.

CAI—"Calc-aluminous inclusion," a microscopic bit of material in a meteorite that is rich in calcium, aluminum, titanium, silicon, and oxygen.

Cassiopeia A—The remnant of a supernova that occurred in 1680 (or possibly a few years earlier).

CCD—Charge coupled-device. A sensitive, solid-state camera used to produce images of the sky.

Central Bureau for Astronomical Telegrams—The office in Cambridge, Massachusetts that is a clearinghouse for bulletins of astronomical interest.

Cepheid variable—A very luminous star that varies in brightness because of periodic expansions and contractions. Cepheids are used as "standard candles" to measure the distances to nearby galaxies.

Chandrasekhar limit—The maximum mass of a stable white dwarf star. The limit is about 1.4 times the mass of the sun.

Circumstellar—The region of space immediately surrounding a star, as opposed to "interstellar," the large volume of space between the stars.

CNO cycle—The process by which stars more massive than about 2 solar masses burn hydrogen to form helium. Carbon, nitrogen, and oxygen act as catalysts, and the relative abundance of nitrogen is enhanced in the process.

Cobalt-56—A radioactive isotope of cobalt resulting from the decay of nickel-56 produced in supernova explosions. Its half-life is 77 days, and it produces gamma rays as it decays.

Common nova—A violent explosion of a star in a binary system caused by the nuclear burning of material on its surface. Common novae usually fade within weeks.

Continuous spectrum—A spectrum unbroken by absorption or emission lines. It is produced by a hot opaque substance, such as a solid or the dense interior of a star.

Crab Nebula—The remnant of the supernova of 1054, in the constellation of Taurus. It is also referred to as M1.

Deflagration—The rapid "burning-away" of a carbon-rich white dwarf in a Type I supernova.

Degeneracy—A state in which particles are packed together so tightly that they can withstand enormous pressures. White dwarfs are supported by degenerate electrons, and neutron stars, by degenerate neutrons.

Doppler shift—The discrepancy between the wavelength emitted by a moving source of light and the wavelength received by an observer.

Ecliptic—The yearly path of the sun around the sky.

Einstein Observatory—A satellite-carried X-ray telescope, which provided images and spectra, that was launched in 1978 and operated until 1981.

Electromagnetic radiation—A form of energy carried by electromagnetic waves. Types of electromagnetic radiation include gamma rays, X rays, ultraviolet, visible, and infrared light, and radio waves.

Electron—A negatively charged particle, found in the region surrounding the nucleus of an atom. Its mass is about 1/2000 that of the proton.

Element—A fundamental chemical substance, characterized by the number of protons in its nucleus.

Elliptical galaxy—A spheroidal or ellipsoidal group of from billions to trillions of stars, with little gas and dust.

Emission line—A bright line in a spectrum. If the spectrum is represented as a graph of intensity versus wavelength, the emission line appears as a peak. It is produced by atoms of hotter gas seen against a cooler background.

Erg—A unit of energy. Approximately the energy expended by an ant crawling over a pebble. A Type II supernova produces 10^{53} ergs of energy.

Exponential decay—A fall in intensity by equal percentages in equal times.

Flocculent spiral galaxy—A galaxy showing a chaotic, random structure in its arms.

Gamma ray—A form of electromagnetic radiation of very short wavelength, less than 0.01 angstrom.

Gaseous nebula—A cloud of interstellar gas, heated to incandescence by nearby stars.

Globular cluster—A spherical group of a million stars or so found in the Milky Way and other galaxies.

Grand design spiral galaxy—A galaxy showing a symmetric, well-defined spiral structure.

Guest star—The Chinese term for a nova.

Half-life—In an exponential decay, the length of time for the intensity to fall by half.

Heavy elements—To astronomers, all the elements heavier than helium.

Helium—An element containing two protons in its nucleus.

Hubble relation—The proportionality between the distance of a galaxy and its redshift.

Hydrogen—An element containing one proton in its nucleus.

Hydrogen burning—A nuclear process by which stars produce energy. Four hydrogen nuclei combine to form one helium nucleus.

IMB—The Illinois–Michigan–Brookhaven neutrino detector, housed in a salt mine beneath Lake Erie.

Infrared—Electromagnetic radiation longer than visible light, of wavelength from about 7000 angstroms to 1 millimeter.

International Ultraviolet Explorer (IUE)—A satellite, launched in 1979, that is used to obtain spectra of objects in the sky at ultraviolet wavelengths.

Interstellar medium—Gas and small solid particles (dust) found in the space between the stars.

Iron-56—A stable nucleus produced in the cores of massive stars late in their lives. It is also produced by the decay of cobalt-56 formed in supernovae.

Isotopes—Forms of an atom containing equal numbers of protons, but different numbers of neutrons.

Kamiokande II—Japanese–American neutrino detector housed in a mine in Kamioka, Japan.

Kepler's nova—The supernova of 1604.

K'o-hsing—The term used to designate a "guest star" or nova in ancient Chinese records.

Large Magellanic Cloud—One of the nearest galaxies to our Milky Way: an irregular galaxy 160,000 light-years distant.

Light curve—A plot of the brightness of a star versus time.

Light echo—Rings or arcs of light around an exploding star caused by light reflecting off of surrounding clouds of dust.

Light-year—The distance light travels in a year. About 6 trillion miles.

Low-mass star—A star whose mass is about that of the sun, or less.

Luminosity—The rate at which energy is emitted by a source. The sun emits 10^{33} ergs of energy each second.

Lunar mansion—One of the 28 divisions of the sky used to indicate position in the Chinese astronomical records.

Magnetic field—The region of space around a magnet where charged particles undergo deviation from straight-line motion. Magnetic fields also exist on planets, stars, and in interstellar space.

Magnitude—The astronomical measure of brightness. A difference of 1 magnitude corresponds to a ratio of brightness of about 2.5. A difference of 5 magnitudes is a ratio of brightness of 100. The planet Venus, at brightest, is about magnitude -4; the limit of the naked eye is about magnitude 6.5.

Main sequence star—A star that is burning hydrogen in its core.

Massive star—A star more than about 8 times the mass of the sun.

Messier catalog—A list of nebulae compiled by Charles Messier in 1771.

Milky Way Galaxy—The spiral galaxy in which our sun is located.

Mystery spot—The companion object 1/20 arc second from SN 1987A discovered using speckle interferometry.

Nebula—An object that looks cloudy or indistinct through a small telescope. For historical reasons, the term is applied to incandescent clouds of gas, to clusters of stars in the Milky Way, and to galaxies outside the Milky Way.

Neutrino—A particle produced in some nuclear reactions. It interacts very weakly with other matter. Its mass is very small—probably zero.

Neutrino reheating—A process in which the "thermal neutrinos" produced by the hot neutron core of a supernova heat the outer layers of the star and blow it away.

Neutron—A particle found in the nuclei of atoms. It carries no electric charge, and its mass is approximately equal to the mass of the proton.

Neutron star—A dense star composed entirely of neutrons. A typical neutron star has a diameter of about 10 kilometers and is about as massive as the sun.

Nickel-56—Radioactive isotope produced in supernova explosions. Its half-life is 6 days, and it emits gamma rays as it decays.

Nova—The term applied to any new star that appears in the sky where none had appeared before. Astronomers often use it loosely to refer to common novae.

Nova Persei 1901—A common nova that appeared in 1901. Astronomers used the light echo from the nova to determine its distance.

Nucleon—A particle in the nucleus: a neutron or proton.

Nucleosynthesis—The process of formation of the elements.

Nucleus—The central core of an atom, containing protons and neutrons.

Orion—A bright constellation visible in the winter sky.

Parallax—The shift in position of an object against a more distant background when viewed from different positions.

Photomultiplier—A sensitive vacuum-tube device used to convert faint light into pulses of electricity.

Photon—A particle of light.

Planetary nebula—A spherical shell of star ejected by a giant star late in its life.

Plerion—A supernova remnant whose interior seems to be filled with material that emits radio waves and X rays.

Polarized light—Light whose electric field is vibrating in one preferred plane.

Presolar nebula—The cloud of gas and dust that contracted to form the solar system.

Proton—A positively charged particle found in the nuclei of atoms. Its mass is approximately equal to the mass of the neutron.

Pulsar—A rapidly spinning, magnetized neutron star that produces regular pulses of radio waves, and sometimes visible light and X rays.

r process—The element formation process in which nuclei capture successive neutrons very rapidly, building up neutron-rich isotopes.

Radioactivity—The emission of particles or gamma rays by a nucleus.

Radio wave—A form of electromagnetic radiation of relatively long wavelength, greater than 1 centimeter.

Red supergiant—A relatively cool star, hundreds of times the diameter of the sun, in the late stages of its life.

S Andromedae—A supernova that appeared in the Andromeda Nebula in 1885.

s process—The element formation process in which nuclei capture successive neutrons slowly enough to undergo beta decay between each capture.

Schmidt telescope—A wide-field telescope used for astronomical survey work.

Schwarzschild radius—The critical size for the formation of a black hole of a given mass. Matter within the Schwarzschild radius can't get out; nor can light.

Sequential star formation—The triggering of new star formation from interstellar gas by the blasts of nearby supernovae.

Sextant—A device for measuring the angles between celestial objects.

Shock wave—A moving region of compression and heating produced by matter moving faster than the speed of sound through a material.

Sk −69 202—The blue supergiant that exploded producing SN 1987A.

SN 1987A—Supernova discovered by Ian Shelton, Oscar Duhalde, and Albert Jones on February 24, 1987.

Solar mass—An amount of mass equal to that in the sun, i.e., about 2×10^{30} kilograms.

Speckle interferometry—Technique used to eliminate the blurring of astronomical images by Earth's atmosphere.

Spectroscopy—The analysis of light by splitting it into its component wavelengths.

Spectrum—The rainbow produced when light is dispersed into its component wavelengths. Often represented as a graph of intensity versus wavelength.

Spiral galaxy—A pinwheel-shaped group of several hundred billion stars. Examples: the Milky Way Galaxy, the Andromeda Nebula.

SS 433—A peculiar object that seems to be either a black hole or a neutron star ejecting powerful jets of gas.

Standard candle—A light source of known luminosity that can be used as a gauge of distance in astronomy.

Stellar wind—A flow of material from the surface of a star into the surrounding space.

Sublunar—In Aristotle's cosmos, the region below the sphere of the moon, where all changes took place.

Supercluster—A cluster of clusters of galaxies.

Supernova—An extremely energetic explosion that destroys a star.

Supernova remnant—Gaseous debris expelled by a supernova explosion. Older remnants include a large percentage of interstellar material swept up by the expelled debris.

Synchrotron radiation—Polarized light and radio waves produced by electrons moving at high speeds in a magnetic field.

Tarantula Nebula—Also known as 30 Doradus. A bright gaseous nebula in the Large Magellanic Cloud.

Taylor-Sedov expansion—Stage of expansion of a supernova remnant after it has begun to sweep up a bit of interstellar material. Characterized by the formation of the hot shock wave that is beginning to decelerate.

Tycho's nova—The supernova of 1572.

Type Ia supernova—A supernova produced by the rapid deflagration of a carbon-rich white dwarf in a binary system. The white dwarf is totally dispersed by the blast.

Type Ib supernova—A supernova produced by the collapse of the core of a star that has lost its outer layer of hydrogen through stellar winds.

Type II supernova—A supernova produced by the collapse of the core of a massive star. May leave behind a neutron star.

Ultraviolet—Electromagnetic radiation of wavelength shorter than visible light, i.e., between about 4000 and 100 angstroms.

Variable star—A star whose brightness varies with time.

VLA—The Very Large Array. A system of 27 radio dishes, located in Socorro, New Mexico, used to obtain very detailed radio images of celestial objects.

Wavelength—The distance between successive crests or troughs of an electromagnetic wave.

White dwarf—A star supported by the pressure of degenerate electrons. Low-mass stars end their lives this way. A typical white dwarf has as much mass as the sun but a diameter about that of Earth.

X ray—Electromagnetic radiation of wavelength shorter than ultraviolet light, i.e., between about 100 and 0.01 angstroms.

Index